ARE YOU SURE?

ARE YOU SURE?

THE UNCONSCIOUS ORIGINS OF CERTAINTY

VIRGINIA "GINGER" CAMPBELL, MD

ACADEMY
PRESS

For permissions requests write to:

Virginia "Ginger" Campbell, MD
c/o *Brain Science*
9340 Helena, RD, Suite F #320
Birmingham, AL, 35244

Ordering information: Quantity and special discounts are available on quantity purchases by educational organizations, corporation, associations, and others. For information contact the address above.

Edited by: Kassandra White
Cover design by: Luigi99 from 99Designs.com
Cover image from iStockphoto.com
Typeset by: Medlar Publishing Solutions Pvt Ltd., India

Printed in the United States of America
ISBN: 978-1-951591-25-0 (paperback)
ISBN: 978-1-951591-26-7 (ebook)

Library of Congress Control Number: 2020905323

First Edition, June 2012, Second Edition, June 2020.

This publication contains the opinions and ideas of its author. It is intended to provide helpful and informative material on the subjects addressed in the publication. It is sold with the understanding that the author and publisher are not engaged in rendering medical, health or any other kind of personal professional services in the book. The material may include information, products, or services by third parties. As such, the Author and Publisher do not assume responsibility or liability for any Third-party material or opinions. Readers are advised to do their own due diligence when it comes to making decisions. The author and publisher specifically disclaim all responsibility for any liability, loss or risk, personal or otherwise, which is incurred as a consequence, directly or indirectly, of the use and application of any of the information contained in this book.

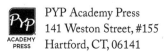 PYP Academy Press
141 Weston Street, #155
Hartford, CT, 06141

DEDICATION

This book is dedicated to the loyal listeners of my podcast
Brain Science (formerly the *Brain Science Podcast*).

While this book is based on several early episodes,
it is because of you that I continue to produce new episodes.
This new edition of *Are You Sure?* would not be possible without
your ongoing support and encouragement.

CONTENTS

PART III
A SKEPTICS GUIDE TO THE LIMITATIONS OF NEUROSCIENCE

ACKNOWLEDGMENTS

This short monograph was inspired by *On Being Certain: Believing You Are Right Even When You're Not* (2008) and *A Skeptic's Guide to the Mind: What Neuroscience Can and Cannot Tell Us about Ourselves* (2013) by Robert Burton. I gratefully acknowledge Dr. Burton's generosity in letting me quote liberally from his books and for allowing me to use the contents of the interviews he did for the *Brain Science* podcast in 2008 and 2013.

I want to thank Kinsey Schwartz, Victoria Pendragon, Daryl Bambic, Gary Issac, Don Gregory, and Dr. Vantipalli Ammineedu for proofreading the first edition of this manuscript. I also want to thank Jeff Moriarty and Evo Terra for helping make the first edition a reality.

Jenn Grace from Publish Your Purpose Press® helped me take this effort to a higher level.

Finally, I want to thank all the unnamed reviewers of this second edition.

PREFACE TO SECOND EDITION

When *Are You Sure? The Unconscious Origins of Certainty* was first published in 2012; it was a minimally edited transcript of two early episodes of the *Brain Science* podcast, which was launched in 2006. Thankfully *Brain Science* is still going strong, and the topic of certainty remains as relevant as ever.

The second edition of this book includes an edited version of a conversation with Dr. Robert Burton about his book, *A Skeptic's Guide to the Mind: What Neuroscience Can and Cannot Tell Us About Ourselves* (2013), which was a follow up to *On Being Certain*.

The overall goal of *Are You Sure?* remains the same. I will give you a practical feeling for the importance of the mounting evidence that much—if not most—of what our brain does is outside both our conscious access and our conscious control.

Virginia "Ginger" Campbell, MD
March 2020

INTRODUCTION

Where were you on September 11, 2001? No doubt you have vivid memories of that horrible day, and you feel certain that your memories are accurate. Where does that certainty come from? Should you trust it?

Are You Sure? The Unconscious Origins of Certainty exposes the unconscious origins of the feeling of certainty. It presents the evidence that the feeling of knowing comes from processes in our brain that are outside our conscious awareness and control. An important goal of this discussion is to consider the implications of this surprising discovery.

Part I is a detailed discussion of the ideas presented in *On Being Certain: Believing You Are Right Even When You're Not* (New York: St. Martin's Press, 2008) by retired neurologist Dr. Robert Burton.

Part II is an edited interview with Dr. Burton that expands on the ideas presented in Part I.

Part III is an edited interview with Dr. Burton about his follow-up book, *A Skeptic's Guide to the Mind: What Neuroscience Can and Cannot Tell Us About Ourselves* (New York: St. Martin's Press, 2013).

This volume is based on three episodes of the *Brain Science* podcast originally posted online in 2008 and 2014. Although they are taken from the past, these ideas remain just as relevant as ever. I want to use this opportunity to thank Dr. Burton for taking the time to talk to me and also for acting as a mentor.

PART I

A LOOK AT THE UNCONSCIOUS ORIGINS OF CERTAINTY

CHAPTER 1

THE FEELING OF CERTAINTY

In recent years, quite a large number of popular books have high-lighted the role of unconscious decision making. In fact, it seems that the more we learn, the more we appreciate that much of what our brains do is beyond our conscious awareness and control. Many of these unconscious processes influence our choices and decisions. Dr. Robert Burton's thought-provoking book *On Being Certain: Believing You Are Right Even When You're Not* (2008) takes a slightly different look at the role of the unconscious. The focus is on what he calls "the feeling of knowing."

In his preface, Dr. Burton opens his book with this line: "Certainty is everywhere." But, after giving numerous examples, he notes that "modern biology is pointing in a different direction" because it challenges "the myth that we know what we know by conscious deliberation." He spends the first part of this book explaining how unconscious parts of our brain create our sense of certainty. He writes, "I have set out to provide a scientific basis for challenging our belief in certainty." He acknowledges that this will also show the limits of scientific knowledge. Dr. Burton goes on to say that his "goal is to strip away the power of certainty by exposing its involuntary neurological roots. If science can shame us into questioning the nature of conviction, we might develop some degree of tolerance and an increased willingness to consider alternative ideas."[1]

The theme of this book is the feeling of knowing and where it comes from. We all know this feeling that Dr. Burton calls "the feeling of knowing." For example, when you feel that you know the answer to a question but you can't think of it right off, that's the feeling of knowing. Similarly, when you're trying to figure out something like an equation and suddenly it all makes perfect sense, that "aha" moment is the feeling of knowing. Of course, sometimes we recognize it more by its absence. To illustrate this, I'm going to quote a paragraph from page 5:

> A newspaper is better than a magazine. A seashore is a better place than the street. At first it is better to run than to walk. You may have to try several times. It takes some skill, but it is easy to learn. Even young children can learn it. Once successful, complications are minimal. Birds seldom get too close. Rain, however, soaks in very fast. Too many people doing the same thing can also cause problems. One needs lots of room. If there are no complications it can be very peaceful. A rock will serve as an anchor. If things break loose from it, however, you will not get a second chance.

What do you think he's talking about in that paragraph? Take a moment to think about this before you go to the next page.

---

[1] Unless otherwise indicated, all quotes are from *On Being Certain: Believing You Are Right Even When You're Not* (New York: St. Martin's Press, 2008) by Robert Burton.

Now, what if I tell you a single word: the word *kite*? Now go back and read that last bit again, and see how much different it seems. For example, the sentence "At first it is better to run than to walk" has an entirely different meaning when you know that we're talking about flying a kite.

A key idea here is that the feeling of knowing makes contemplating alternatives difficult. Can you go back to reading this paragraph as you did before I gave you the word *kite*? Burton's book *On Being Certain* explores the neural underpinnings of this feeling of knowing.

How do we know what we know? There can be a mismatch between what we consciously think we know and what we actually know. Burton starts exploring this principle by asking his readers to remember where they were either when President Kennedy was assassinated, when the *Challenger* exploded, or when the attacks of 9/11 happened. Think back on those three events. Perhaps you only remember 9/11. The strange thing is that people will have very different memories about where they were during these events, even if they compare their memories with those of people they were actually with during the event.

There's a study called the *Challenger* study[2] that demonstrates this. This study was done by a scientist named Ulric Neisser, who was studying people's memories of and about highly dramatic events. Within a day of the *Challenger* explosion, he interviewed 106 students, and he had them write down exactly how they heard about it, where they were, what they were doing, and how they felt. Two-and-a-half years later, he interviewed them again, and he found that for 25 percent of them their second account was significantly different from their original journal entries. In fact, more than half the people had some degree of error, and fewer than 10 percent gave all the details exactly the same as they had originally.

---

[2]Ulric Neisser and Nicole Harsh, "Phantom Flashbulbs: False Recollections of Hearing the News About the Challenger," in *Emory Symposia in Cognition, 4: Affect and Accuracy in Recall: Studies of "Flashbulb" Memories*, ed. Eugene Winograd and Ulric Neisser (Cambridge: Cambridge University Press), 9–31. Now available for free at https://psycnet.apa.org/record/1993-97049-001.

Before they saw their original journal entries, most of them were certain that their memories were absolutely correct. In fact, many of them, when confronted with what they had originally written, still had a high degree of confidence in their false recollections—despite being faced with journal entries in their own handwriting because they just felt that their current memories were correct. In fact, there was one student who said, "That's my handwriting, but that's not what happened." And lest you think that this was an isolated incident, there have been plenty of other documented cases. You can test this yourself, as Burton did when he asked several of his medical school classmates where they were on the day Kennedy was assassinated, which occurred while he was a medical student. When he conducted this research, he got a similar reaction. As this *Challenger* study shows, people consciously choose a false belief because it feels correct.

This conscious choice is an example of what is called cognitive dissonance, which was a term coined back in 1957 by Stanford professor Leon Festinger. He defined *cognitive dissonance*[3] as a distressing mental state during which people "find themselves doing things that don't fit with what they know or holding opinions that do not fit with other opinions that they hold."[4] The key idea is that the more committed we are to a belief, the harder it is for us to relinquish it, even in the face of overwhelming evidence. We tend to go with what *feels* right, no matter what the evidence. I'm sure that if you think about it, you could think of an example of this from your own life.

So what causes the feeling of certainty? What could be the advantage of an unjustified *feeling of knowing*? Where does this feeling of certainty come from? There's pretty good evidence that it must be coming from someplace in the brain, as there are some medical conditions in which

---

[3] For an excellent description of cognitive dissonance, I highly recommend *Mistakes Were Made (but Not by Me) Why We Justify Foolish Beliefs, Bad Decisions, and Hurtful Acts*, rev. and new ed. (New York: Houghton Mifflin Harcourt, 2020) by Carol Tavris and Elliot Aronson.

[4] Festinger, Leon. *A Theory of Cognitive Dissonance*. Stanford: Stanford University, 1957.

the feeling of knowing gets distorted. For example, Capgras syndrome is a condition in which people feel that their loved one has been replaced by an impostor. This is because they have lost the feeling of knowing that loved one—even though they can logically say, "Yeah, it looks exactly like them." Sometimes, this involves something like a piece of furniture that they think has been replaced, and they'll say, "Yeah, it has all the exact qualities, but it's just not the same piece of furniture." This condition has been documented in several patients. Remarkably, patients don't have the same type of brain lesion, but what they do have in common is that they all tend to choose in favor of what they feel. Based solely on their feelings, they come up with tortured logic to support their conclusions.

This leads to the question: why does our physiology seem to be weighted in favor of feeling over logic? Apparently, conviction isn't really a choice. As Burton points out in his book, studies of blind-sight demonstrate that knowledge and the awareness of knowledge arise from different regions in the brain. In the case of blind-sight, people have damaged the part of the brain that allows them to have conscious awareness of vision; however, visual information goes to other parts of their brain, and they are able to act in ways that show that this visual information is actually reaching those other areas, even though they have no conscious awareness of being able to see. In *On Being Certain*, Dr. Burton also considers other examples, such as the physiology behind mystical experiences and the effects of certain drugs.

What about the emotional element of certainty? This seems to be rooted in older parts of the brain, including the cingulate gyrus, amygdala, hippocampus, hypothalamus, and basal forebrain structures. These are the areas that have, in the past, been called the limbic system and are associated with emotions and drives. For example, when discussing the emotion of fear, scientists will look to the amygdala.

Joseph LeDoux is well-known for figuring out the role of the amygdala in fear. He did experiments with rats during which he paired a sound and a shock. These experiments showed that there is an acoustic pathway that actually bypasses the auditory cortex. LeDoux showed that if you cut the acoustic nerve, there will be no fear response. But, if you remove the auditory cortex itself, it doesn't have any effect on the fear response. In other words, the rat doesn't have to be aware of the sound to feel fear. This established that the amygdala has a role in the fear response that does not require any conscious awareness or recognition of the stimulus. In a subsequent experiment, researchers showed that if the amygdala is destroyed, the animals are totally fearless, which implies that the fear response really *does* require a functioning amygdala.[5]

Antonio Damasio has done a lot of work with emotions in human patients, and he has argued that normal emotions, including fear, are essential for normal decision making. Similarly, it has been discovered that seizures involving some of these so-called limbic structures can cause things like a sense of something familiar or déjà vu or the sensation that things are totally unreal so that they feel like a dream. Burton points out that he is not claiming that the limbic system is the source of our sense of knowing. His point is that our sense that something is familiar or real is *not* a conscious conclusion because it's coming from lower parts of the brain that we cannot control or consciously access.

Does that mean that the feeling of knowing is an emotion? Well, Burton quotes Antonio Damasio as saying "deciding what constitutes an emotion is no easy task."[6] Something like fear might seem straightforward, but if we look at something like gratitude, which we would generally consider an emotion, we don't really observe that gratitude disappears with any particular known brain lesion, nor can gratitude be elicited by stimulating a certain part of the brain. Another example is surprise, which is not easily elicited by stimulating the brain.

---

[5] These results have since been challenged because of cases that appear to show that some people are able to experience fear even in the absence of the amygdala. LeDoux himself now claims that the amygdala is part of a survival circuit and that fear is learned. This is a very complicated subject—beyond the scope of the current discussion.

[6] Antonio Damasio, *The Feeling of What Happens: Body and Emotions in the Making of Consciousness* (New York: Houghton Mifflin, 1999), 66.

So how should we approach this feeling of knowing? Burton concludes that we should regard mental states as *sensations*. He says this reflects the physiological truth that they come from the relatively discrete output of local neural structures, just like vision comes from the sensory output from the eye. In his later work, he makes this parallel more explicit by using the term *mental sensations* to capture the essential idea that experiences, such as the feeling of knowing, are generated automatically by unconscious processes, just like our perception of sensations such as sight and sound.

Regarding mental states as sensations recognizes that they are subject to the same principles as other sensations. For example, if you cut a nerve to your finger, you can't will it *not* to feel numb. People who have obsessive-compulsive disorder, for example, can't will the feelings that drive their compulsions to disappear. We can't will ourselves to have the feeling of knowing or to get rid of it when the feeling contradicts what appears to be logical. The key idea is that just as sensations, such as vision, really aren't under our conscious control, neither are many mental states.

On page 40 of *On Being Certain*, Burton makes this bold but important claim: "By using these criteria of universality, relatively discrete anatomical localization, and easy reproducibility without conscious cognitive input, the feeling of knowing and its kindred feelings should be considered as primary as the states of fear and anger. At this point, we may not be able to figure out exactly where in the brain the feeling of knowing is generated, but we definitely have overwhelming evidence that it is *not* under conscious control" (italics added).

CHAPTER 2

THE BRAIN'S HIDDEN LAYER

To sum up where we are so far, we've reached the conclusion, based on the evidence, that the feeling of knowing is a primary mental state. It is not dependent on any underlying state of conscious knowledge. This conclusion is important because it determines how we might try to model the role of certainty in the hierarchical organization of the brain—as if it is similar to a neural network. To do this we also have to understand the key elements that govern the brain's hierarchical structure. Finally, when we consider how the contradictory aspects of thought collide, we'll be able to see why Dr. Burton says, "Certainty is contrary to basic biological principles."

Looking at the brain from the point of view of neural networks, he points out some basic facts. First, at the level of the neurons, the key structure is the space between the neurons, which is called the synapse. At a synapse, there are many[7] different neurotransmitters involved in neuronal activation. The dendrites receive many inputs, and they also participate in feedback loops. In the end, a neuron either fires or doesn't. A key idea here is that when the neuron fires, *you lose all the information that was combined*

---

[7] This estimate is now greater than one hundred.

to trigger its firing. (This is just a small example of the information that is lost before anything reaches our conscious awareness!)

Two important concepts here are the idea that there is a hidden layer that's inaccessible to our consciousness and that this inaccessibility cannot be overcome. To illustrate the concept of a hidden layer, consider what happens when you buy a book from Amazon. If you've ever done this, you know that the website automatically recommends other books that you might like. Obviously, it uses some algorithm to generate its suggestions, but you don't have any access to how it did that. This algorithm is an example of a hidden layer. In the case of consciousness, the irony is that even though we don't know how consciousness occurs, we know from a conceptual standpoint that it has to arise from a hidden layer of neurons. You cannot tell from the output what happened in the hidden layer. This is important, and I will come back to it.

First, we need to consider the principles of modularity and emergence. Burton defines a module as a cluster of highly individualized neurons devoted to a specific function. For example, there are at least thirty discrete modules devoted to vision. Some of these have been discovered just because of what happens when they're damaged. On the other hand, if you look at individual neurons, they're pretty similar no matter what part of the brain you look at. There are many different kinds, but there is no such thing as a "super neuron." So we need the concept of emergence, which he describes on page 59 of *On Being Certain*: "Consciousness, intentionality, purpose, and meaning all emerge from the interconnections between billions of neurons that do not contain these elements." He concludes that "[m]odularity combined with a schematic hierarchical arrangement of increasingly complex layers of neural networks and the concept of emergence serves as an excellent working model for how the brain builds up complex perceptions, thoughts, and behaviors."

I think that the easiest way to think of emergence is the old saying "the whole is more than the sum of its parts." However, when it comes to the idea of modularity, Burton warns that if you try to apply it to behavior, it can lead to excessive reductionism because there aren't really well-defined, discrete collections of neurons committed to behavior as there are for things such as vision.

With behavior, what we're looking at are widely separated but interconnected neurons. So I guess that would make them more like functional modules in some sense but not necessarily discrete modules. For example, learning to read[8] involves a number of different cortical centers that probably first evolved for other types of visual recognition.

Dr. Burton concludes that the feeling of knowing is universal and probably generated by a local region in the brain since it can be generated by direct stimulation or chemical manipulation. We have already discussed the important fact that we don't have conscious control over it. However, as far as I know, the precise localization of where this comes from has not yet been discovered.

---

[8] Maryanne Wolf, *Proust and the Squid: The Story and Science of the Reading Brain* (New York: Harper, 2007). This book was discussed in episode 24 of the *Brain Science* podcast, and Dr. Wolf was interviewed in episode 29. Both episodes are available online at http://brainsciencepodcast.com/.

CHAPTER 3

THOUGHT AS PERCEPTION

S o far, we have claimed that the feeling of knowing is generated in a hidden layer of the brain in the sense that it is not accessible to our consciousness or to our control. So that brings us to the following questions: When does a thought begin? and When it comes to the feeling of knowing, what is its relationship to thought?

Dr. Burton writes about several different scenarios. One scenario would be when you have the feeling of knowing *without* an accompanying thought, such as during a mystical experience. Another scenario would be when the feeling of knowing and the thought reach conscious awareness simultaneously, as in the "aha" moment. Then there's the situation during which you encounter an idea and objectively determine it to be correct, such as when you're looking for your friend's house and he answers the door, so you know for sure that you're at the right house. Whether or not we should unconditionally trust our feeling of knowing really depends on which scenario has occurred.

Why do we get the feeling of knowing without proof? There are a lot of situations where we have to act *without* sufficient information. For example, when a baseball player hits a fastball, it has been proved that the player has to start his swing before he can actually pick up the path of the ball. Yet players feel absolutely sure that they saw the ball, even though it is physiologically impossible.

Burton says, "If the brain did not somehow compensate and project the image of the approaching ball back in time, you would see the ball approach after you hit it." He says that the window of time required to process sensory data before it is perceived is what allows the brain to create a seamless world of now. Research indicates that these backward projections can be as long as 120 milliseconds. On page 75, he says, "[t]he basic neurophysiological principle is that the need for immediate response time reduces the accuracy of perception of incoming information."

He also mentions that even though most people don't engage in high-speed sports, we do the same sort of thing when we engage in everyday speech. We are usually anticipating what people are going to say before they say it; because we tend to hear what we expect, we also fill in little gaps and ignore repeated words. This may be one reason why it is so hard to create accurate speech recognition with computers.

With an example such as the baseball player thinking he saw the ball that he could not have seen, we can understand how the subjective backward referral of the feeling of knowing can lead to erroneous conclusions. These sorts of temporal illusions can also occur over much longer time spans. The question *when does a thought begin?* May be impossible to answer scientifically because thoughts are inaccessible to standard scientific measure *until* they reach consciousness. It might not be easy easy to determine the relationship between the feeling of knowing and our thoughts. The key idea here is that our internal brain time may not be an accurate reflection of external time. This is an area I explore further in Part II.

Here is another practical example of how our brains trick us: I like to play tennis, and one of the ongoing controversies in tennis is line calls. When you're playing a recreational-level match, you don't have line judges, and your opponent basically calls the lines on your shots to his or her side. One of the problems is that in reality, you can't see the ball bounce. Instead, your brain takes the information about the speed of the ball and its trajectory and then creates the impression in your mind that you have seen it bounce. Of course, you think you have seen it bounce, even though experiments repeatedly show that it is physiologically impossible for humans to detect anything happening that quickly. The point is that the feeling that you saw the ball bounce is entirely convincing. This is an example of how the brain makes us believe that it knows things that it can't really know.

Now let's return to the issue of people having different memories of shared events. First, I need to point out the difference between episodic and semantic memory. Episodic memory is the memory for the events that one experiences, and semantic memory is the memory for facts. Episodic memories are very dynamic. They change over time according to what else we experience. But even if we know this, none of us believe that our memories are as fragile as they really are. As Burton says, "[w]e cling to our belief that our pasts approximately correspond to our memory." If you're interested in learning more about how our memories can be so inaccurate, you might want to look at the work of Daniel Schacter, including his book *The Seven Sins of Memory*.

Burton suggests that we make a distinction between those thoughts that arise from the complex computations of hidden layers of neurons, as in episodic memory and semantic thoughts, which require the memorization of facts. The former kinds of thoughts are constantly changing. He calls these "perceptual thoughts" because he thinks that trying to call them "episodic thoughts" would be cumbersome.

In his terminology, "perceptual thoughts" refers to the thoughts that result from complex computations outside of our conscious awareness. Using the word *perceptual* reminds us that there can be illusions. We know that we can have illusions of our visual perception, so by using the term *perceptual thoughts*, Burton emphasizes that the processes going on outside of our conscious awareness can not only change our thoughts, but they can also mislead us.

CHAPTER 4

WHY IT FEELS GOOD TO KNOW YOU ARE RIGHT

Earlier I asked the question, What would be the possible benefit of a feeling of knowing that is actually false? This brings us to a consideration of our brain's reward systems and how they interact and influence our thoughts. We know that there are extensive connections between the pleasure reward systems, emotions, and the opioid peptides in the brain. Research suggests that the mesolimbic dopamine system is a key component of the brain reward circuitry. It originates in the upper brainstem and connects to parts of the brain that are involved in emotion and cognition, including parts of the frontal lobes and the nucleus accumbens, which is thought to be involved in addiction. Scientists have shown that brain-mediated rewards cause behaviors, including addictions, to persist.

So, you have to wonder, how does this relate to the feeling of knowing? Dr. Burton gives the example in *On Being Certain* of a person who, when faced with a charging lion, climbs up a tree and survives. After the person escapes, he has the feeling that he has learned something. If you make these sorts of decisions repeatedly, you will probably have a positive feeling of correctness, which becomes linked to that behavior. Dr. Burton argues that the feeling of knowing and feelings of familiarity are integral to learning.

Think about it. When we look at the world around us, what is the first thing we notice? We notice if anything has changed, which means that we're comparing what we see to what we expect. That is why Dr. Burton calls these feelings "thought's original 'yes man.'" Early on, this feeling of being right was associated with simple choices whereas now, we have thoughts that are much more complex, and we often face questions that have no clear-cut answers. In fact, often we will never know for sure if our choice was the right one. Yet our brain's reward system still requires a signal! We have to feel that the thought is worth pursuing before we have any supporting evidence or justification. If we were focused on the uncertainty, we'd be stuck or paralyzed.

It seems reasonable to assume that our brain's reward system, including the part that relates to the feeling of knowing, represents the result of evolutionary adaptations. Burton warns that these sorts of explanations may be overly simplistic because we really don't have any way of knowing what the original adaptation must have been. Somewhere along the line, our brains, as he puts it, "stumbled" across the potential for abstract thought, but there had to have been some sort of appropriate reward system.

What keeps you going when you're working on a long-term project? We have to have some sort of conviction that our efforts are worthwhile. However, whatever psychological motive we invoke doesn't really address the underlying physiology of how our brains reward this kind of behavior. Burton proposes that the feeling of knowing was already part of the feedback reward system for learning. He says, "An unwarranted feeling of knowing *might* serve a positive evolutionary role." Note that he says "might." He's not really making any unprovable claims.

Obviously, the reward has to be strong enough to keep us going until our thoughts are verified. So that means that the feeling of knowing has to feel very similar to how we feel when we actually do know something for sure. Burton suggests that there's a spectrum of bridging motivations, ranging from hunches, gut feelings, faith, belief, and profound certainty, and these all contribute to the feeling of knowing. The bottom line is that we need some sort of reward, so the neural connections binding the thought and the sensation of being correct is gradually strengthened.

This strengthening happens with repetition, and once the connections are formed, we know that they're difficult to undo. For example, Joseph LeDoux has shown with rats how persistent both fear responses and addictions can be. Thus, once established, emotional habits and patterns of expectation and rewards are difficult to eradicate. Dr. Burton argues that the same principle applies to thought, so once we have a thought connected to a feeling of correctness, it's hard to undo. He speculates on whether an insistence on being right might be similar physiologically to other sorts of addictions. He writes, "Might the know-it-all personality trait be seen as an addiction to the pleasure of the feeling of knowing?"

In light of what we know about brain plasticity, I wonder whether black-and-white thinking, such as the way schools emphasize the right answer, could mold the brain's reward system to prefer certainty over open-mindedness. Another element is the fact that just as people experience different levels of pleasure from things such as alcohol, they might also experience differing levels of pleasure from the feeling of knowing. Our ability to accept uncertainty may reflect both our experience and our genetic makeup. Dr. Burton's conclusion is "[a]ny present-day understanding of how we know what we know must take into consideration the contradictory nature of thought's reward systems. The feeling of knowing, the reward for both proven and unproven thoughts, is learning's best friend and mental flexibility's worst enemy.[9]"

To repeat, the feeling of knowing is learning's best friend and mental flexibility's worst enemy.

In *On Being Certain*, Dr. Burton tries to dispel the misconception that because we each have what he calls an "innate sense of reason," we should be able to overcome our perceptual differences and see problems from some sort of optimal perspective. But what about the genetic component of the feeling of knowing? For example, twin studies have shown that

---

[9] Burton (2008) page 101.

genes appear to affect our interest in religion and spirituality. Of course, as he reminds us, human behavior is exceedingly complex, and you can't look at it as just a product of genes. However, in terms of the genetic component of behavior, the amygdala seems like a reasonable place to start.

Scientists have learned how to create what they call "knockout" strains of mice. In a "knockout" strain they knock out a gene for a particular thing and see what happens. For example, they have developed a strain of mice that lacks the ability to make stathmin, which is a protein that is normally found in high levels in the amygdala. They found that these "knockout" mice are very difficult to condition to fearful stimuli.[10] It is interesting to speculate how such an important adaptive response could be affected by just one gene.

But, as Burton points out, there's an important difference between innate tendencies and predicting actual behaviors. For example, in the twin studies, it was found that what people said they wanted to do and what they did were not the same. As he says in the book, "[d]esire and action are not synonymous." It also seems that our apparently deliberate reasons for making a particular decision will be influenced by our innate risk tolerance. Dr. Burton doesn't really think we could ever sort out the genetic component of something as complex as the feeling of knowing.

Then, there's the fact that the environment actually affects gene expression. For example, we know that our brains become biased to hear the sounds that we are exposed to when we are young. The auditory cortex actually gets tuned to those sounds. For example, Michael Merzenich[11]

---

[10] Gleb P. Shumyatsky, Gaël Malleret, Ryong-Moon Shin, Shuichi Takizawa, Keith Tully, Evgeny Tsvetkov, Stanislav S. Zakharenko, Jamie Joseph, Svetlana Vronskaya, DeQi Yin, et al., "Stathmin, A Gene Enriched in the Amygdala, Controls Both Learned and Innate Fear," *Cell* 123, no. 4 (2005): 697–709, https://doi.org/10.1016/j.cell.2005.08.038: "The knockout mice also exhibit decreased memory in amygdala-dependent fear conditioning and fail to recognize danger in innately aversive environments."

[11] Li I. Zhang, Shaowen Bao, and Michael M. Merzenich, "Persistent and Specific Influences of Early Acoustic Environments on Primary Auditory Cortex," *Nature Neuroscience* 4 (2001): 1123–30.

has shown with young rats that this actually starts from the very beginning of brain development. Thus, nurture affects nature due to the fact that
experience actually affects gene expression. Hopefully, this knowledge will
give us a new understanding of why we should not expect to be able to get
others to think and believe as we do.

Now most of us have learned that this is true, but we keep thinking it
should be otherwise. The point here is that this is the way our brains are! It's
physiologically impossible that we will all be thinking the same thoughts.
As Dr. Burton says on page 123, "To expect that we can get others to think
as we do is to believe that we can overcome the innate differences that
make our thought processes as unique as our own fingerprints." This is a
very important idea. It fits in with one of Dr. Burton's main themes, which
is that understanding how our brains really work in this regard ought to
encourage us to try to have more tolerance for different beliefs.

CHAPTER 5

THE UNCONSCIOUS ISN'T ALWAYS RIGHT

I have written about how thoughts can have a perceptual quality to them, which includes the fact that they are partially generated by things that we can't control.

Next, we're going to consider the idea of sensational thoughts. What does this mean? It relates to the relationship between our thoughts and the fact that we are embodied. The way we have been thinking about thought and reason really challenges a concept we've inherited from the ancient Greeks, which is that "reason" exists as something objective. Dr. Burton quotes the famous linguist George Lakoff here as saying, "Reason is not disembodied as the tradition has largely held, but arises from the nature of our brains, bodies, and bodily experience. Reason is not a transcendent feature of the universe or a disembodied mind."[12]

Dr. Burton says on page 127, "Disembodied thought is not a physiological option. Neither is a purely rational mind free from bodily and mental sensations and perceptions." You can't have a thought without sensation. For example, you have a sensation that you're inside your body or inside

---

[12] Dr. Burton takes this quote from George Lakoff and Mark Johnson, *Philosophy in the Flesh: The Embodied Mind and Its Challenge to Western Thought* (New York: Basic Books, 1999), 4.

your head. This is created by your brain and your body. We can't localize where the sense of self comes from, but we do know that the right temporal lobe is important because if you stimulate certain areas in the right temporal lobe, you can cause a person to have an out-of-body experience. This implies that our sense of being *in* our body is created by our brain.

Our awareness that we are thinking is also a sensation. Thoughts that don't reach awareness don't feel like they're being actively thought. For example, let's take the idea of sleeping on a problem. Let's say we have a problem, and we say, "I'm going to sleep on it." We go to bed, and we wake up the next morning with the answer. When we wake up, it feels like the answer just came to us, and somehow, that feels different than if we had gone through a deliberate, conscious process. If we remember something this way, it also *feels* different, even though it really isn't any different as far as our brain is concerned. There is a disconnect between what we know and what we feel.

This brings us to the subject of unconscious thoughts. Several recent books have touted the importance of the unconscious in decision making. However, Burton challenges the idea that we should regard these unconscious decisions as correct or special. Suppose you program a computer to solve a problem and you go off to do something else. Later you come back and the computer has generated an answer. You don't look at that as being some sort of miracle or product of intuition just because it came about outside of your awareness. Yet if you talk to, say, a writer who has let an idea percolate for a bit while he's working on something else, when the solution comes into his consciousness, he's very likely to attribute it to something special like his muse or a higher power. This idea that creativity, for example, comes from some special mystical place is very common. These ideas that seem to just "pop into our head" seem to have a special sense of rightness (feeling of knowing) that arises at the same time as the idea reaches our awareness.

Burton suggests that we have difficulty assigning intention to thoughts that occur outside of consciousness. It doesn't feel intentional without some sort of clear and immediate preceding effort. This is where we come back to the example of the computer. We don't need to think the computer has an intention to solve the problem. But when it comes to our own thinking, on page 133, he says, "Any significant delay between a question and an answer tends to strip the thought of a sense of being intentional." This is probably related to the fact that we seem to be wired to see cause as preceding effect.

Recall the example of the batter starting his swing before he actually sees that path of the ball. Neural mechanisms make him feel that he saw the ball before he began his swing. This cause-and-effect way of perceiving the world seems to be the way we are wired.

Now, Burton is not suggesting that the feeling of knowing is present in the unconscious. In fact, he says, "An unfelt feeling makes no sense." What he is suggesting is that it is likely that there's some sort of unconscious pattern recognition going on, which contains a calculation of the probability of correctness. Then, when it reaches a certain level, we consciously experience a feeling of knowing. That is, the closer the fit is between a previously learned pattern and the pattern that comes in, the greater the feeling of correctness.

A common example is when we can't think of someone's name, but we say, "I don't remember her name, but I'll know it when I hear it."

We don't know how thoughts emerge from neurons, but I agree with Burton that both conscious and unconscious thoughts are likely to be rooted in the same mechanisms, which are basically neural networks processing information. This fits what we already know about how the cortex processes things, such as sight and sound. So basically, he is saying that thoughts are happening in the same way. Otherwise, we would have to postulate a difference between conscious and unconscious thoughts, which would mean that the basic biology of cognition changes when a thought moves in and out of consciousness. Now that might make sense from some philosophical point of view, but it doesn't really make any sense based on what we know so far about how the brain works.

Rejecting what he calls "the dubious premise" that unconscious thoughts represent a different way of thinking, Burton asks, "[Why not consider cognition as a single entity that isn't subdivided into various ways of being experienced?" After considering how this approach might be applied to popular psychology terms such as *intuition* and *gut feelings*, he emphasizes that recognition and discussions of these sensations of a thought are integral to any theory of mind. As he points out, thoughts require sensory information. In fact, our brain even has sensory systems that selectively tell us when we're thinking a thought. This explains how we experience cause and effect, intentionality, and the feeling of knowing or its absence.

The key idea is that we know the nature and quality of our thoughts via feelings, not reason.

According to Burton, on page 139, "feelings such as certainty, conviction, rightness, wrongness, clarity, and faith arise out of involuntary mental sensory systems that are integral and inseparable components of the thoughts they qualify." This has some very important consequences. For one thing, the findings of neuroscience really challenge our ideals of reason and objectivity. As he says on page 141, "certainty is not a biologically justifiable state of mind. There is no isolated circuit in the brain that can engage in thought that is free from involuntary and undetectable influences."

Objective thought is not humanly possible!

So how do we find a way to portray the mind that's both emotionally satisfying and reflects these limits? Can we move beyond what Burton calls "the myth of the autonomous, rational mind," which is the belief that we can step back from our thoughts in order to judge them? Trying to ignore the evidence is not the answer. When examining ideas about the mind, he suggests that we need to ask whether the idea is a personal opinion or something from pop culture or a scientific hypothesis. Most important, always ask, "Is this idea consistent with how the mind works?"

Burton devotes quite a bit of time to considering two recent best sellers: *Emotional Intelligence* and *Blink*[13] and he criticizes some of the authors'

---

[13] Daniel Goleman, *Emotional Intelligence: Why It Can Matter More than IQ* (New York: Bantam Books, 1995); Malcolm Gladwell, *Blink: The Power of Thinking without Thinking* (New York: Little, Brown, 2005).

conclusions. The biggest thing that he challenges is the idea that we can get to these layers and, therefore, be able to judge when they're right and when they're leading us astray. Instead, he feels that we need to go back to the position of Timothy Wilson of the University of Virginia, who said, "Our minds have evolved to operate largely outside of consciousness."[14] Wilson's position was we can't know our unconscious. If that's the case, we really can't make our unconscious smarter, and we can't regard our unconscious as having some part that is totally logical because we really can't get to it to make those judgments.

Now, without unconscious cognition there would be no conscious decision making, but Burton argues that we do not have any criteria for knowing when we can trust our split-second decisions. He points out that feeling satisfied *later* about a decision is not a valid criterion for deciding that it was definitely the right choice. The feeling that a decision is right is not the same thing as providing evidence that it is right.

This brings us back to the myth of the autonomous, rational mind and objectivity. The point is that neither one of these is possible. Think about the famous experiment during which people were told to watch a video and count how many times a basketball was tossed back and forth. A guy in a gorilla suit walks by, and they don't even see him, maybe because they were so focused and what they saw was determined by what they were looking for.[15] Burton says it should be a red flag whenever anyone claims that they had no a priori assumptions. He concludes that complete objectivity is not possible. The best we can hope for is partial objectivity, which means acknowledging our biases and assumptions as accurately and honestly as we can. For example, when you look at a good scientific experiment, the researchers tell you what their assumptions were. It is essential that we remember that we cannot access the unconscious through introspection. We can only get a partial view of how our minds work. No matter how

---

[14] Timothy D. Wilson, *Strangers to Ourselves: Discovering the Adaptive Unconscious* (Cambridge, MA: Belknap Press, 2002), 16.

[15] For more information, see *The Invisible Gorilla* website at http://theinvisiblegorilla.com.

much you meditate or do things to develop more awareness, you cannot get to these unconscious layers.

Why is it so difficult to let go of this myth of the autonomous, rational mind? One reason is that we generally tend to feel that the mind is in a different category from the body. We accept our physical limitations, but we somehow think we should be able to overcome our mental limitations. Burton observes that "our mental limitations prevent us from accepting our mental limitations."

CHAPTER 6

IN PRAISE OF TOLERANCE

I want to summarize what we've talked about so far. First of all, I think everyone would agree that we need to have a feeling of knowing to function. But the important principle to grasp is that the brain creates some sensations that *feel* like thought but really aren't. Emotional components, like the feeling of knowing, can't be separated from "pure" thoughts. In fact, I believe that there's no such thing as a pure thought. Thus, we can think that the feeling of knowing is a response to a thought when actually the feeling came before the thought. This reflects our brain's ability to distort our sense of time.

Also, and this is very important, we can't dismiss the feeling of knowing, partly because we don't have any access to how it's created. The most important thing that you need to remember from this discussion is that the feeling of knowing is coming from parts of the brain that we cannot access or control.

So how are we going to make practical use of this knowledge? The first suggestion Dr. Burton makes is that we should try to replace the word *know* with *believe*. For example, he suggests something like "I *believe* that the theory of evolution is correct because of the overwhelming evidence." He says, "Substituting 'believe' for 'know' doesn't negate scientific knowledge. It only shifts it from being unequivocal to highly likely." I think this is what science ought to be about anyway because we're supposed to always be open to new information.

Dr. Burton is optimistic that this knowledge about where the feeling of knowing comes from could restore our tolerance for conflicting opinions. He also says that we ought to take the point of view, and I agree with him, that any idea that isn't capable of being independently tested should be considered a personal vision. He wants psychology to abandon the idea of a perfectly rational unconscious and let go of the idea that we can know when to trust our unconscious.

The bottom line is that certainty is biologically impossible. Dr. Burton believes that we can learn to tolerate the contradictory aspects of our biology and that we can learn to tolerate the unpleasantness of uncertainty—perhaps even teach this ability. Modern science incorporates the language and tools of probability and statistics, which make uncertainty more manageable. In real life, we constantly make decisions with incomplete information. But we also seem to have the tendency to feel certain about these choices.

PART II

INTERVIEW WITH ROBERT BURTON, MD

One of the really wonderful things about the mind is that it feels different—that's where this age-old body-mind thing comes from is that in the very big picture we feel the mind to be separate from the rest of what the brain does, because that's what the brain does.

—Robert A. Burton, MD

2008

CHAPTER 7

INTRODUCING ROBERT BURTON, MD

Note: This interview has been edited for clarity. The full interview is available for download at http://brainsciencepodcast.com. Look for the *Brain Science* podcast, episode 43.

Dr. Campbell () and Dr. Burton (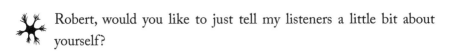)

Robert, would you like to just tell my listeners a little bit about yourself?

Sure. I graduated from Yale, then I went to med school at UCSF [University of California, San Francisco]. I stayed there for a hundred years, completing my internship and residency before staying on at UCSF to run the neurology department at one of its hospitals, Mt. Zion. One day, I got the writing bug and started writing some novels, with my first novel coming out in '91. Actually, *On Being Certain*, started as a novel, but I realized I couldn't do it in the novel format because it was better suited to a nonfiction book on science and philosophy.

 So how did you come to write *On Being Certain*?

I was going to write a novel about brain stimulation, and at the same time, I was also collecting observations on a variety of neurological phenomena, such as Cotard's Syndrome—where patients believe they are dead. Even when you show them that they have a pulse, they can't believe they're not dead. One patient, who had encephalitis, was particularly striking. When I showed her that she had a pulse, she said, "Well, this only proves that dead people can have a pulse."

Fortunately, my agent said, "These are really interesting, but what's the point?" So I tried to figure out what had prompted me to gather together these particular case histories. Eventually, it occurred to me that the basis of this book was "how we know what we know." It morphed from novel to memoir to its present form.

 Did you have a particular goal in mind for what you want the book to accomplish?

 No, at heart I'm a novelist. In a novel, you start a character in motion, walk him down the street or put him in a chair. You make readers want to see what happens to him. In this case, I tried to find out what these neurological observations had in common. It wasn't until after I realized that the feeling of knowing was an involuntary sensation that I then went back and redid the whole book from that point of view. Only then did I recognize my overarching point: certainty or feelings of conviction or rightness aren't within our conscious control. One of the reasons the book took so long to gel was that I never heard anybody discuss mental sensations as a separate category of human experience. It wasn't until that occurred to me that this all fell into place.

 Well that probably is why you were able to come up with an original way of looking at this. The reason I devoted two episodes[16] to your

---

[16] Episodes 42 and 43 of the *Brain Science* podcast are available at http://brainscience podcast.com/.

book was that I think it is important to share this idea that the feeling of knowing is not under conscious control.

I suppose so. I talked to a philosophy professor at Berkley who read my book, and he said, "You know we don't talk about this in philosophy." Of course, nobody in neurology talks about these sort of neural phenomena because they're not something that you can prove.

I do wish to emphasize, probably more than anything else, that I think the luxury of not having to do this for career advancement allows you to make mistakes and take false turns. It allows you the time to think about something long enough and eventually you might even come up with something original—because you're not forced to prove something. I really didn't know what my purpose was nor did I feel any urgency to know what it was. In writing a novel, you don't have to have a point or know the purpose behind what you're writing. I think it makes it a lot easier than when you're forced by circumstances to write to try to prove something.

CHAPTER 8

THOUGHT AS PERCEPTION REVISITED

I came away with the feeling (no pun intended) that one of the important concepts in your book is the perceptual nature of thought, looking at thought more like the way we look at vision. We understand that optical illusions trick our visual perception, but we somehow think that our "thinking" should not be prone to such things.

Correct, and I think most of us accept that the brain does everything we experience in terms of thoughts and feelings and all the rest. But we don't somehow recognize that these all come out of little cells that then connect together, and they make a higher level of group. So, let's say you take individual cells that see one aspect of motion; one will see lines, and another will see borders, and they all merge together to make an elemental image of something. Then, another will recognize color, or location or motion. Finally, there'll be another higher level that generates an image that somehow goes into consciousness and allows a person to see a butterfly.

I don't think we recognize that one of the things that the brain generates is the sensation of the mind looking down at the butterfly. However, this sensation comes out of the same material, so that the mind itself is generated by the same neurons that are generating the rest of the

39

perceptions. All of these neurons, collectively in neural networks, are subject to all sorts of perceptual illusions. One of the really wonderful things about the Mind is that it feels different. This age-old mind-body thing really stems from the fact that in the very big picture we feel the Mind to be separate from the rest of whatever the brain does because that's what the brain does.

It's a perceptual illusion at the very core. So all these debates about mind–body really are just a misunderstanding about how the brain works.

 Didn't you say something about how our mental limitations make it impossible for us to accept our mental limitations?

Right. We use the sensations from our brain to determine what our brain is doing, but we can't see our brain or the unconscious, so the limitation we really have is that we can only see what our brain tells us we can see. When our brains tell us about what we can see, it's saying to you, "Oh, I'm thinking about this." All these questions about free will and stuff really get down to what you experience at a conscious level versus what you're actually doing.

One of the things that relates to that is when the ideas seem to come out of nowhere because we don't have access to where the thoughts came from before they reached our conscious awareness, and we get all kinds of distortions about when this happens. Would you like to talk a little bit about how our brain distorts our perception of *when* things happen?

Sure. The example I use in the book is straightforward—imagine that you go to your friend Joe Blow's house and you sit in his living room. He's there, and you make a very strong mental association that he lives on Filbert Street, let's say, and you recognize the front porch and the house and so on. That's how you learn where his house is. His house is not only a visual image, but it's now embedded with the sensation,

"Yes, that is his house" because that's the only way you would learn it. So, that feeling of knowing that we talked about really is about the likelihood of correctness of the visual image you see. Now twenty years later, he invites you over, and you're driving down the street. The only way that you could possibly recognize his house is to already know that it's his house. But it wouldn't make sense for you to look at the houses and think, "Ah, that's his house" until after you had seen it. The way your mind experiences this is, "Let's see—I'll drive down here and ... red house, white house, blue house. Oh! There's the house with the porch—that must be Joe's house." The fact that you knew it was Joe's house made you recognize it in the first place, so you knew, in advance, but felt the sensation of knowing after the fact.

This rearrangement of time is essential for learning something and being able to recognize anything. For example, if I recognized you in the street, my calculation would have been, "There's very high likelihood it's Ginger Campbell." Then I would say, "Oh, that's Ginger." However, it would have felt like the calculation occurred after the fact. We just don't know when thoughts begin.

I don't even know the initial impetus for this book. I do know one thing; after my mom passed away, I found a term paper in her collection of stuff that she'd left for me, and it was a college paper I'd written on the same subject forty years ago. I had no recollection of the paper, which I'd written when I was seventeen or something. Obviously, it had been percolating along in the unconscious for decades. Why it suddenly emerged into something I had to write is a complete mystery. You asked me when I started this book. I could say a few years ago, but in reality, it probably began in my unconscious several decades ago.

When you have an idea that you aren't consciously thinking about, say a problem in your novel, or whatever, and the solution comes to you; it seems common for people to attribute that to some kind of mystical source instead of the fact that it actually came from their own brain.

Correct. And you know it's interesting, I wrote one novel where I could not think of the ending, and really, it went on for like five months. Nothing worked. Then one night, I remember someone was telling me about a friend who had a cardiac arrest in Bangkok. I remember thinking about it and going to bed. In the middle of the night, I wakened, and the whole thing appeared. So I jumped up and wrote it down, and it stayed verbatim except for the usual editing. I thought, "Wow, how did that happen?"

This shouldn't be a surprise when you think about thought as being primarily unconscious. Your unconscious is simply waiting for the piece that makes all the other pieces fall into place. Your unconscious is primed and ready to go; all it needs is this last bit of information. For me this last piece sent the ending into consciousness, along with the idea that the ending was "right." It wasn't some muse or some extraterrestrial being that sent this information to me; it was just the missing piece from my unconscious, but that's not how it felt.

CHAPTER 9

FEELING RIGHT IS REWARDING

I'd like to ask you about how all this ties into the reward systems of the brain. What do you think the key idea is about how this feeling of knowing relates to the reward system of the brain?

I went through a series of thought experiments to try and figure this out. One of the questions is, Why would you pursue a thought which you had no idea was correct? In other words, if you envision the unconscious as generating virtually an infinitude of ideas—some of which will rise up into consciousness, many of which will go by the boards—then the question may be "Why does something arise in the consciousness?" Then you say, I'm going to think about this. Well, there has to be some unconscious mechanism generating some idea of the likelihood that this is something worth pursuing. Whether it's aesthetic, revising a scientific discovery, or going to make you money, whatever it is, there'd have to be some reward for your unconscious thought.

What would it be? Well, it would have to be some sense that pursuing the idea will ultimately result in something worthwhile. So there'd have to be a good feeling or sense of motivation. Otherwise, every single thought that your unconscious mind had would appear in consciousness—irrespective of its value. Therefore, I thought, there's got to be some feeling or some quality that arises that makes some thoughts preferable to others, prior to you knowing they're any good. There's got to be something that feels good about the thought. A good example is the hunch. You wake up in the morning and go, "I've got a great idea." Now what you really have is a feeling that you've got a great idea, along with the idea. You have no idea if it's a great idea.

So you really do need a reward system for thinking, and it's probably as powerful as for everything else. If you don't have it, you won't do anything. When you see a kid who's stuck in high school algebra or Latin or whatever it is, and he goes, "I don't know why I have to do this"—you can't get him to do it until somehow he understands maybe he's going to get into college with it—or whatever the reward may be—but if he doesn't understand, he won't do it. Most thoughts we have we have because they feel worthwhile. So the reward system has to be really prime. I think now the evidence is the connections between the limbic system and the brain; the mesolimbic reward systems are really quite powerful, and I think that you can't even think of a thought without a reward.

 You also brought up the idea that maybe some people might actually be addicted to feeling that they're right all the time.

Well, you know I try to strike a middle ground in genetics because I don't think genes control everything, but they obviously have a significant influence on how you think. I'm on the far end of the skeptical continuum; I just don't get that feeling of knowing very often. So when I hear an opinion, I tend to wonder if the opposite is true! On the other hand, we all know the people that are always right; they cannot be wrong, and they seem to get a great deal of joy out of being right. I wouldn't be surprised if it is a form of addiction.

Most of us are kind of in the middle: do you think that we can learn to be more comfortable with uncertainty, since the reality is that most of the time we have to make a lot of decisions without being really sure?

Well, I guess that also gets down to the pleasure, which gets back to your reward thing; I actually enjoy being uncertain, and I enjoy the idea of ambiguity. It's probably one of the reasons I like writing novels. I like the idea that there can be multiple solutions or even multiple endings! There can be multiple things a character can do—all that would be believable for that character. I've never really enjoyed starting with the premise "I am going to prove this." I remember sitting down with one of the really famous best-selling mystery writers, and he said to me, "I always write the ending first because I want to know where I'm going." I said to him, "If I knew where I was going, I wouldn't start the book."

Exactly.

I do believe that there are actually different ways of thinking about the world that come out of your biology. You ask me if you can teach people to be comfortable with uncertainty, but I don't know that you can teach people, who love to be right, not to be so sure of themselves or to allow room for other options. You would have to start really young and train them to understand that it isn't necessary—or even possible—to be right all the time. Now, this might be an illusion, the idea that you're correct, but once the circuitry's firmly in place, it's like having a bad backhand in tennis; a learned habit is really hard to get rid of.

I know in medicine it seems like people in different specialties have different personalities; I'm an emergency room doctor, and we're always having to act with incomplete information, and then the radiologists drive us bonkers, you know, because they're always sending us reports that say, "Well, it might be this or it might be that—order another test," which does us no good …[17]

 I started in neurology, in part because they didn't have any tests. When I started my residency, they didn't have CT [computerized tomography] scans and MRIs [magnetic resonance images] …

It was all physical exam …

And history. It was fun to speculate, and it was also fun because there was a lot of leeway for error. You knew you were going to be wrong a fair percentage of the time because you had no way to confirm your ideas. This attracted the people who were sort of like gentlemen philosopher doctors. Now with the genetic code, all of the molecular biology, et cetera, you're attracting a different breed of person. So I really do think not only the fields you go into, but even what branch of medicine you choose is in part dictated by how you feel about uncertainty and not having to be correct.

Do you want to say anything else about where genetics comes in? I agree with you that we can't ever really explain something as complex as behavior purely with genetics, but when it comes to this feeling of knowing, do you think there's anything else we need to say about that?

---

[17] In emergency situations, we often don't have time for more tests. Our first priority is determining whether our patient has a life-threatening condition. More tests don't make patients better. Even in an office setting, the first priority is having enough information to choose an initial treatment. A skilled physician only orders the tests needed to treat the patient. Unfortunately, this pragmatic approach is being lost because patients increasingly seem to equate lots of tests with good care. I would contend that the opposite is often true! More tests should happen, only if they are really needed, such as when the patient is not getting better or if a preliminary test suggests a dangerous condition.

One thing. When you start a thought, you feel as though you know how a thought begins. For example, you're trying to decide whether or not more oil drilling in Alaska is a good thing. We think our thoughts are going to be a rational exploration of exploration of oil, OK? Well, try a simple experiment. If I say to you, "*Exxon Valdez,*" what's the first image you get?

Oil spill.

Right. Now, someone else might say "long gas lines."

If you ask people what image they first imagined, you will immediately see that they divide themselves into two groups that are coming to the question from different starting points.

I have a family member who said "long gas lines" because he doesn't have the same starting point for thoughts that then generate his thinking about oil drilling. I think in part it depends a lot upon how your memory is stored. You mentioned on the podcast about the mice that, by knocking out a single gene, no longer store fearful memories. Now imagine a politician with that genetic variation. He's not going to be fearful of an oil spill in the future because recollections of past oil spills don't trigger any painful memories. He isn't fearful of the consequences and would be more likely to think that technology can solve any serious ecological problem. Though he feels that his reasoning begins with conscious deliberation about oil drilling, it actually begins with an unconsciously generated image along with a lack of sense of fear that the oil spill generates in those of us with sad, fearful memories of the *Valdez* disaster.

So, I think that genes influence your thoughts that lead to different behaviors, and no two people think exactly alike. If somebody who is fearful about an oil spill talks to someone who isn't, they're not starting on a level playing field.

CHAPTER 10

THE MYTH OF THE AUTONOMOUS RATIONAL MIND

I want to talk about what you have called "the myth of the autonomous rational mind" in a minute, but we talked about thought as being a perceptual thing, and you also talk quite a bit in the book about thought as a sensation. Those seem to be like overlapping ideas, but what does it mean to look at thought as a sensation?

I don't know if thought is a sensation, but our feeling about a thought has to be. For example, when I was finishing up the book, I was in bed with my wife, and I said to her, "What was the name of that cartoon figure that was the possum that was all philosophical?" She couldn't remember, and I said, "Watch this. I am going to do an experiment on my mind, in which I'm going to intentionally ask my unconscious: who wrote that comic strip and what was the name of the comic strip." So I went to bed, and when I woke up in the morning, sure enough it was Walt Kelly and *Pogo*. The answer came right up, and I said, "Oh, see it did work—just like that."

Now, the important thing to notice is, though I could not feel my mind thinking, it was clearly working to come up with the name.

49

The only way I was aware of the answer was because I was aware that suddenly the answer appeared in my mind, but I was never aware of thinking of it.

So now you ask, "Well, what's the difference between that kind of thought and the thought that you're doing consciously?" Well, one aspect, which is certain, is that you feel like you're thinking. And I've never heard it described, but the only way you know you're thinking consciously versus unconsciously is the sensation of the thought.

So, when I say the thought is a sensation, at least in part conscious thought has to be a sensory awareness of what the brain is doing, as opposed to what the brain is doing alone. If it was just what the brain was doing, I would have been aware of all the information for *Pogo*. I never got all the names of the various cartoon characters I went through to find that name. So, it has to have a sensation for us to know we're doing it. I think that this is another thing that's been sorely overlooked, which that the brain has ways to notify itself of what it's doing, so it's up to the unconscious to tell you when you're thinking.

Just like we're not aware of all the stuff that goes on before what we consciously see reaches our awareness, all that processing, we don't see that either.

That's exactly correct. There was a study done recently where some neuroscientists did these fMRI [functional magnetic resonance imaging] studies, in which they asked people to press buttons with either their left hand or their right hand while they were in a scanner. The decision was completely arbitrary. It turned out that at least 5 seconds before they made any conscious choice, the appropriate area of the brain lit up on the MRI scan. The unanswerable question is who chose what? In other words, from a research perspective, the choice of which hand to use was decided unconsciously and then came into consciousness, but from a conscious perspective you feel like you chose it sometime later. This gets back to that same thing—perception of time. If you look at it that way, there are sensory phenomena within the brain that dictate how we feel about our thoughts and actually guide them. Does that make sense?

 Yeah. One of those subjects that's been going on, on my discussion forum[18] has been how this relates to the idea of free will, and one person wrote, "Well, if it was in my unconscious it was still me."

That's right, and that's one of the major questions. Most of us only think of our consciousness as being our "self," so to speak. Then there is this whole diminished capacity argument about, well, he couldn't help himself because his unconscious told him to do it. I think the idea of what is "you" is an outdated term, based upon the idea that we are conscious, rational human beings. Once you realize that a lot of what we do is unconscious, then the words free will or decision making need to be changed to choice. The unconscious has all kinds of choices because it's making them all the time. Free will implies that it's free of material influence, which is probably a non sequitur, probably meaningless …

Your brain is materialistic at some level, and yet you do have choices. I had a choice not to do this interview or to do this interview. So yes, we do have free will, but it's of a different sort—it's more of a choice. Ben Libet[19]—I don't know if you are familiar with him …

Yes.

He and I shared adjacent offices at the Neuroscience Institute at Mt. Zion, and he used to tell me, "You know, your unconscious makes all the decisions, but your conscious still retains the veto power."

I think this is a profound comment because he seems to be implying that even though your idea appears to you from out of the unconscious, you can then contemplate it, and you can input your conscious thoughts back

---

[18] The discussion forum can be found at http://brainscienceforum.com.

[19] Benjamin Libet, *Mind Time* (Cambridge, MA. Harvard University Press, 2004).

into the unconscious and change the way you think about things. There is a feedback loop from the conscious to the unconscious.

 Which I think means that in the end, we have responsibility for the choices that we do make.

 Correct. I totally agree, and it's not an all or nothing; there's intermediate ground; we have choices within the limits of our biology.

CHAPTER 11

COGNITIVE DISSONANCE

 Would you talk a little bit about the idea of cognitive dissonance?

Right. Cognitive dissonance. Obviously, everybody understands it in the psychological terms of "Oh, how could he believe that when all the evidence is to the contrary?" For example, if you take somebody who believes physically in the resurrection; he has to know at some level that this isn't physiologically possible as we understand biology. Yet, it has great meaning for him in terms of his iconography for religion. Well, under these circumstances, it would seem good, up until recently, to make some sort of psychological discussion of why he might feel that way; maybe, he wants to be religious, maybe; he's got deep-seated needs, et cetera, but the fact of the matter is that even though it manifests as a belief system, the main thing is that the feeling of knowing is so strong that even if he knows the physiology doesn't make sense, that feeling that "the physiology doesn't make sense" is not as strong as the feeling of "Yes, this occurred."

You know there are lots of people who just simply cannot accept life without flying saucers or UFOs or alien abductions … there's lots of so-called cognitive dissonance that I actually think emanates from an overpowering feeling of knowing on a physiological basis. As to why that occurs, maybe, in part it is psychological, maybe, in part it's because of people's predisposition to, or the ease with which they can experience this feeling. I actually don't know why some people are more prone to it than others, but it ends up being a physiological phenomenon at the root of cognitive dissonance, namely an inappropriate feeling of knowing making a wrong idea seem more correct than evidence to the contrary.[20]

One of the things you said several times in *On Being Certain* was that, in general, the feeling of knowing tends to overrule the logical or the evidence for most people, even when they're confronted with the facts. Why? Why does the feeling win out?

Because, I think at the heart of it all we are emotional beings. Just take falling in love. You've never heard of anybody who's been dissuaded by cool reason from attaching themselves to somebody who, from an outside perspective, is a horrible choice. They might even agree with your arguments, but the emotion of love overpowers whatever comes second. I guess it has strong evolutionary roots. I mean, I'm not sure whether it's got to do with procreation, but love clearly outdistances anything that comes second, such as reason. Right? By the same token, if you didn't have a feeling of knowing, there'd be no reason to do anything. You'd be totally lost. The feeling of knowing is very powerful and without it you would have no learning. In other words, the only way you can really learn is by, "2 plus 2 is 4; isn't that right?" And the teacher says "Yes, that's right Johnny, you did fine," and eventually, you feel really good about 2 + 2 = 4, and you learn that. Well, if you didn't have that feeling, you would learn nothing.

So I think that this feeling of knowing must be enormously powerful; it must subvert reason when the two come into conflict. One is more

---

[20] For an excellent exploration of cognitive dissonance, I recommend *Mistakes Were Made (But Not by Me): Why We Justify Foolish Beliefs, Bad Decisions, and Hurtful Acts*, rev. and new ed. (New York: Houghton Mifflin, 2020), by Carol Tavris and Elliot Aronson. Tavris appears in episode 43 of *Books and Ideas*, which is available at http://virginiacampbellmd. com.

biologically necessary. I think reason came much later and is much less of an evolutionary necessity than this feeling of correctness, which is at the heart of learning, at the heart of what causes reasoning in the first place. The reasoning is dependent on the feeling; the feeling isn't dependent on the reason.

 Kind of brings us way back to what David Hume wrote. I think he argued that the mind was really the stepchild of our emotions.

Correct. That is correct. You cannot have reason without that feeling because you would never be able to go, "Oh, I think that's right." On the other hand, you can certainly have the feeling without the reason. That's what the "aha" is—that sort of "aha" where life is wonderful. You don't need the reason for that.

CHAPTER 12

SHOULD YOU TRUST YOUR INTUITION?

I really appreciated the fact that on several occasions in your book you argued for more tolerance, from both ends—from believers and nonbelievers toward each other's stances on these things, since we really don't have conscious control over when we have that feeling. Just like you said you're on the skeptical end. That's not something that you can will.

No, it really isn't, and it's just the way I am. You know I got an email the other day from a prominent trial attorney in America who said he's writing a book for lawyers, and after reading my book, he's writing a chapter on how to view the opposing lawyer, the jury, and the judge. Because of my book, he feels you should say, "Well each of these people has independent lines of reasoning arising out of dissimilar unconscious mechanisms, such as the hidden layer we talked about." Then you have to look upon them with more respect and remember they aren't just "dodos" that disagree with you. I thought this was really interesting because all of a sudden he saw a jury box as twelve different hidden layers, as opposed to a group of people that you had to convince you were right.

That's a great example. You had a chapter where you talked about gut feelings and intuition, and you particularly talked about the book *Blink*,[21] which I bring it up because it's a book I have talked about in the past. Do you want to say anything about that before I comment?

Well you know, I don't mean to do character assassination because Gladwell's a great writer. He is a sensationally good *New-Yorker*-style writer, and I admire him greatly. But I think in this particular case he's made not only a faulty assumption but one which is really dangerous. He has said that he wrote his book based in part upon reading a book by Timothy Wilson, *Strangers to Ourselves*[22], which is about the unacceptability of unconscious thought. Gladwell thought, well maybe there are ways that you can train this in some way. So he uses examples where he claims unconscious thought turns out to be as "right" as more deliberate thought. But the fact of the matter is, all thought originates in the unconscious. It is then modified, scrutinized, examined, and tested in the conscious. So the idea that the unconscious is suddenly a spring or a marvelous fountain of ideas is a truism. It has no meaning, but the question is, can you know if these thoughts are correct? He uses an example at the start of his book of the Greek kouros,[23] which is a 500 BC statue that was bought for several million dollars. At first the people at the museum thought it was great, but some people came along and said that they felt revulsion at the sight of it, and they felt certain that it was a forgery.

According to Gladwell, it turned out, that these people, were "absolutely right"; by their gut revulsion, these people were able to tell that this was a forgery. No! These people were art history experts who had assimilated expertise into their unconscious thinking processes. If you

---

[21] *Blink: The Power of Thinking Without Thinking* (New York: Little, Brown, 2006) by Malcolm Gladwell was discussed in episode 13 of the *Brain Science* podcast, which is available at http://brainsciencepodcast.com.

[22] *Strangers to Ourselves: Discovering the Adaptive Unconscious* by Timothy Wilson (Cambridge, MA. Harvard University Press, 2002).

[23] For more information about the Greek kouros, see http://en.wikipedia.org/wiki/Getty_kouros.

learn something and you do it over and over again, it becomes part of your unconscious. On an expert level, they made an unconscious decision they thought was right. Well the answer is, they made an expert decision, and it *could* have been right. But it's up to the conscious mind to test it out, to use some sort of empirical method to see whether this really was a forgery.

The fact is that they went out and did all these studies, and they still don't know. I looked this up a few weeks ago in the catalogue from the Getty museum, and they list it as either a 550 BC statue or a forgery; they said it's perhaps because we don't understand the methods of Greek burial statues, so when Gladwell says they were "absolutely right," he's absolutely wrong. In making such an all-or-nothing claim, he's glorifying or deifying the unconscious as having the ability to know when it's correct. The unconscious came up with an opinion, but it didn't come up with scientific evidence. It's up to the conscious person to objectively assess the statue via whatever testing is available, to see if they can justify the opinion. So he can't "know" that this was absolutely right; for him to say that is no different than relying on a physician who says, "I have a gut feeling that you got this or that or the other thing …" No, that's not a proper way to practice medicine. So, I take great exception with his book on that basis.

The idea that came to my mind as I was reading was that he doesn't seem to really make a distinction between the expert knowledge that physicians have acquired, and which has become part of our unconscious circuitry in some ways so that we don't have to think about it every time we see a patient, and the stuff that comes out of our unconscious that's really not based on anything like that.

Yes, that's correct. Take a surgeon for example, or take yourself in an emergency room situation. So, you've had that feeling, "oh, this is nothing …" or "this is a person who may have appendicitis." You make some qualitative judgment. OK, so your experience that you have acquired over a number of years becomes part of your unconscious mental processing. It becomes additional information; it feels as though it's better. For some, medical experience it clearly is better. Obviously, for you, the real criteria would be to do a study of ER docs to see whether or not they're better than chance alone at diagnosing appendicitis based on their clinical experience. If they are, then you've shown statistically that your expertise is of value. In this case you can find out because you can see a lot of people with appendicitis and can correlate your clinical findings with objective findings on CT scans or by directly looking at the diseased appendix. You can judge where expertise works. But in other cases, you can't know for sure whether your expertise is on the money. The Getty is still unsure about the statue because they don't have sufficiently accurate dating techniques.

I think that one of the key things you were trying to bring out in that chapter was that we can't know when the feeling of knowing is right, so we can't stop there.

Right. It's really pretty simple to say, "OK, is this an idea that that I can test, or is this an idea that I have to carry with me as feeling right but I can't yet know if it's right?" That process can sort out the scientific questions from the metaphysical or political or practical. Clearly, you can't know whether it's better to go to the beach or to the mountains for a vacation. One feels better than another, but that's not really the kind of opinion that you can test. You'd have to *know* that. Yet when people say with certainty that we should be in Iraq or we should be out of Iraq, or we should do oil drilling, they act as though these are ideas that can be tested, rather than admitting that they are, at best, personal opinions based upon one's general understanding of the world—but not certainties.

Right. ... Another thing you said in the book was that the feeling afterwards that you're happy with the choice you made is not a valid criterion for deciding it was the right choice.

I mentioned a study[24] in the book that is often cited as evidence that people are happier with their split-second decisions than with decisions they deliberate over. But if your choice is to buy a home in Malibu and it feels great, and then it slides into the ocean, it wasn't a good choice. The real question is, What would constitute the proper criteria for studying good versus bad choices? I don't know what they might be. It's certainly not the fact that it feels like a good choice. Saying a choice is good because it feels like a good choice is a circular argument.

And it brings us back to the whole idea of tolerance, since, as you pointed out, there are an awful lot of things that we can never really know. Stuff that people really care a lot about.

Writing this book took a long time; I have no idea if I'm more tolerant, but I do recognize that other people's opinions arise out of separate lines of reasoning than mine and that in nonscientific matters there is no litmus test to know with certainty who is correct. This knowledge should make me more tolerant; I hope it does! I'm aware of the fact that they're not starting from the same position. Just that alone is helpful.

And I'd like to return to "the myth of the autonomous rational mind" because the fact that this is not really under our conscious control has some big implications, doesn't it?

Absolutely. We are a complex organism that has developed an additional faculty, which is awareness of what the unconscious is doing, but this doesn't make us rational. Instead, this awareness should give us empathy for our biology.

---

[24] A. Dijkesteruhuis, M. W. Bos, L. F. Nordgren, and R. B. Van Baaren, "On Making the Right Choice: The Deliberation-Without-Attention Effect," *Science* 311, no. 5763 (2006): 1005–1007.

In contrast, Richard Dawkins says something about faith as a cop out. Now, I am a 100 percent in favor of Dawkins's ideas on evolution and all the rest of it, but this is like telling someone who's in love, "Don't be in love." That doesn't make sense. Don't have faith. Well, we are all pursuing our biology to greater or lesser degrees, as best we can. I think we need to develop a society that's more consistent with our biologic principles, rather than having some Enlightenment sense that there's a rational mind that should govern the way we are. It reminds me of Nancy Reagan saying, "Just say no to drugs." Well, it sounds great. I have no objection to that, except it just doesn't work. It is foolish to draw up a policy like that because it goes against your biology.

CHAPTER 13

UNANSWERED QUESTIONS

It's been great talking about the ideas in *On Being Certain: Believing You Are Right Even When You're Not*. Would you like to talk about what you are working on now?

Well, I have been thinking a lot about the fact that most of us feel that we are what our conscious mind tells us we are. I think that we sense only what we sense. We sense ourselves as starting an idea, having a thought, making plans, etc., etc. But we really don't see that that facility arises out of the same cognitive stew that causes all these perceptual illusions in general. If you were to take a look at Western thought, it really is all about the mind-body thing. All the major questions arise out of the conception that the mind is somehow a separate entity. The philosopher John Searle wrote a book called *The Mind*, in which he covers all the various kinds of theories, but none of them make any sense if you think about it. Because way down deep, the mind is simply a higher level function that we can't conceptualize, just the same way as you said it's more than the sum of its parts. This whole idea of emergence is really impossible to visually see. In other words, you realize that if you take a chocolate chip and you take some flour and water, there is no embedded cake in there. There's nothing;

there's just chocolate and flour, but we know you can make a cake. Well, the cake is material. You can still see it.

But in this case, the problem is that what the brain generates is immaterial.[25] We can't see it, yet it does exist. It exists in the same way that pain exists. Pain isn't anywhere. The brain doesn't experience pain in the neurons. When you stub your toe, there's no neuron that goes "ouch." It occurs at a higher level. The problem I haven't figured out yet is that there needs to be a metaphor for understanding higher level function when seen from a lower level that will allow people to get rid of this distinction and argument about the mind and free will and causation. These are all problems of language that arise out of misconceptions about what the mind is. I think that might be my next project.

 You pointed out in your book that we don't think pain is something mystical or magical just because it can't be localized (it's emergent), yet somehow it seems natural to look at our mind and feel that it has to be somehow different.

Right. There must be some analogy or metaphor where you can say, well just as pain exists but is undetectable, the mind exists, but it exists arising out of stuff that we cannot control. Even though it feels like it's separate and also feels like it's in control in some sense, it also feels like it's you and feels like a self. These are all phenomena that are undetectable but necessary. I don't want to use the word illusion because illusion implies it doesn't exist, but on the other hand it is an illusion if you mean by illusion you can't see or taste or smell or touch it. It's an illusion without being an illusion.

It's an illusion in the sense that it's not exactly what we think it is.

---

[25] This problem is tackled in great detail in *Incomplete Nature: How Mind Emerged from Matter* (New York: W. W. Norton, 2012) by Terrence W. Deacon.

Right. I think the problem is that most of us feel very, very strongly that we exist as individuals. If you didn't have that feeling, you wouldn't do anything; there would be no motivation. Without a sense of self, there'd be nothing we could refer to as life. Yet, we must recognize that this exists somehow in a different realm, and I use the word 'hierarchy' in the book to go up to the level of consciousness, but I never actually discuss the mind because I don't believe the philosophy we've used has gotten us anywhere with these questions of free will, responsibility, and diminished capacity.

For example, take the case of Robert Alton Harris. I think he's the guy that was the murderer who, after he killed these people, ate their hamburgers. He was sent to the electric chair, but the argument was that he had diminished capacity because he had childhood abuse. In fact, he did have brain damage, but the question is, who is Robert Alton Harris? Is he just a conscious person? Or does he represent all of the biology and abuse? Of course, he represents all of the above, so we need a new way of thinking about what is a person in order to attribute responsibility. If somebody doesn't like reading because they read slowly, we don't punish them if we say they have dyslexia. On the other hand, is it right to tell them not to try to read? We need to know who the whole person is. I suppose as we go along we will also need a whole new conception of what the mind is.[26] So that's my new project.

I heard Michael Gazzaniga[27] interviewed on *All in the Mind* in relationship to this, and he made a good point that when it comes to

---

[26]The implications of neuroscience to the judicial system are examined in David Eagleman's *Incognito: The Secret Lives of the Brain* (New York: Vintage Books, 2011). Dr. Eagleman was interviewed in episode 75 of the *Brain Science* podcast, which is available at http://brainsciencepodcast.com/.

[27]Gazzaniga discusses this in more detail in his latest book *Who's in Charge?: Free Will and the Science of the Brain* (New York: Ecco, 2012). This book is discussed in episode 82 of the *Brain Science* podcast.

these kinds of social things it's also about the interrelationship with the person and his society. In other words, a person needs to know that in the society he lives in it's not OK to kill other people. Gazzaniga thinks that this should be part of the equation when we're deciding about peoples' responsibility for the things that they choose to do.

When I was young, I had the good fortune of meeting David Bohm, a physicist who used to work with Einstein, and he wrote a book called *Thought as a System*. I spent a weekend with him, and I remember him trying to explain what he thought the mind was. He had the idea that it was a different order or different level or emergent level, and he used a metaphor, which I couldn't quite grasp, but he made the point that even our thoughts aren't just ours. In other words, if you watch a bunch of skinny little models and you believe that's the norm, you will inject into your unconscious hidden layer an idea of what the norm is. So when you look in the mirror, how you see yourself will be partly determined by your cultural exposure. When patients with anorexia look in a mirror, they literally perceive themselves as fat even when their retinas "see" a thin person. Your perception actually came out of a social idea, which became your thought.

So David Bohm actually believes that you don't own your thoughts apart from a larger community, which I think is also a fascinating idea.

That was somewhat similar to what Gazzaniga said. Well, I think we're just about out of time—is there anything else you'd like to share before we close?

No, I very much enjoyed chatting with you. I was thrilled that you took the time to read my book; the way you described it, it sounded like it would be fun and interesting to read.

Even though I've written this book and spent a lot of time on it, it always strikes me every time I hear these ideas. They are counterintuitive, yet, at some level, it makes you say to yourself, "that's interesting." It's hard to avoid the fact that your mind reverts back to the default position that this isn't true.

CLOSING THOUGHTS FROM THE FIRST EDITION

One subject that I mentioned early on, which we forgot to come back to, was what are the implications for psychology and psychiatry of the fact that the unconscious really isn't totally accessible, controllable, or necessarily reliable—because obviously there are schools of psychoanalysis that depend on the belief that our unconscious contains truths that are not in our conscious mind. Dr. Burton really doesn't talk about this in his book. He is very conscientious about staying in his own areas of expertise, and he never claims to have any expertise in psychiatry. After all, he was a neurologist, but I think if you're familiar with some of the ideas of psychology, you will see the implications of this information.

The most important concept that Dr. Burton points out is that the *feeling of certainty* about anything that comes out of the unconscious mind is *not the most reliable indicator of correctness*. Then, there is the fact that research is showing that memory is highly dynamic and that it is really easy to create false memories. That, too, has big implications for approaches in psychology that are based on recovering a person's repressed memories about their past.

Just like Dr. Burton, I can't claim to be qualified to draw final conclusions about the implications of this information, but it seems to me that at the very least we should take the approach that information that comes from the unconscious, in any sort of psychoanalysis, should be put to the test. It shouldn't be assumed to be true, and if it strongly conflicts with objective evidence, the unconscious contents shouldn't be, as Dr. Burton called it, "deified." For people who do work in the field of psychology, I think this demonstrates the importance of being aware of modern neuroscience and what it's uncovering and of being willing to give up models of the mind that are not supported by the evidence.

What about the rest of us? As I observed at the beginning of this monograph, we all experience the feeling of knowing. It comes to us in many forms, ranging from simple hunches to our confidence in hard-earned knowledge. In light of the convincing evidence that this feeling of knowing is one of the many things our brain produces outside our conscious awareness and control, we need to remind ourselves to examine this feeling critically whenever it arises.

If we get into the habit of asking ourselves what kind of evidence supports our certainty, we may be surprised how often the strength of our feeling is not supported. Unfortunately, the strength of our conviction is not a reliable indicator that we are right.

At the same time, it is important to remember that we don't decide what we feel. Because the feeling of certainty arises from the unconscious processes of our brain, we can't choose what we feel sure about. Just as we can't choose who we love or don't love, we don't really choose our most deeply felt beliefs. This should help us to be more tolerant both of ourselves and of others.

PART III

A SKEPTICS GUIDE TO THE LIMITATIONS OF NEUROSCIENCE

INTRODUCTION TO PART III

I n 2013, Dr. Robert Burton published *A Skeptic's Guide to the Mind: What Neuroscience Can and Cannot Tell Us About Ourselves*, which is a follow-up to *On Being Certain*. This section contains an edited version of the interview that originally appeared as episode 96 of my *Brain Science* podcast (now called *Brain Science*).

CHAPTER 14

ROBERT BURTON RETURNS

In 2013, Dr. Robert Burton published another thought-provoking book titled *A Skeptic's Guide to the Mind: What Neuroscience Can and Cannot Tell Us About Ourselves*. Since it was a natural follow-up to *On Being Certain*, I was eager to have him back on the *Brain Science* podcast. When I observed that it had been over four years since our first conversation he quipped, "it takes me that long to have a new idea."

What follows is an edited version of our conversation.[28] The full version is available at http://brainsciencepodcast.com.

Dr. Campbell () and Dr. Burton ()

INTERVIEW

 Will you start by giving us an overview of your new book.

 Basically, in *On Being Certain*, my idea was to explore the involuntary roots of the concept of certainty. To do that, after years of

---

[28] We talked about *The Skeptic's Guide to the Mind* in episode 96 of *Brain Science* podcast, which originally aired in April 2013 and is available at http://www.brainsciencepodcast.

thinking about it, I came up with the idea that the brain does its cognitive processing (which we'll call, for the future, just "calculations" or "computations") at an unconscious level, but it also has to have some way of notifying you of the likelihood that it's right.

For example, if you see a face in a crowd, your pattern recognition will give you some idea of whether it's Sam, or Joe, or nobody in particular, to the extent that if it is Sam, you will also have a sensation of familiarity that it's Sam and a sense that your decision on an unconscious level that it's Sam is, in fact, correct.

So, you have really two aspects to any thought: one is the calculation itself, which is devoid of any feeling tone, and some independent assessment of the likelihood that it's correct. So, in *On Being Certain*, I wrote about the involuntary origin of the feelings of being right, and of certainty, and so on. But after having written the book, at the end of the book I had some sort of speculations that were half-formed thoughts in my mind; over the next few years, I've been considering what they meant.

I've come to realize that virtually all of the feelings that constitute a sense of a mind are actually involuntary sensations that are accompanying various aspects of unconscious brain computation. Those not only include feelings of certainty, but also feelings of agency and causation— perhaps even some moral considerations of right and wrong. But they are sensations that arise as an accompaniment of a calculation that your brain is doing primarily—though not necessarily exclusively—at an unconscious level.

 And that's what you called "mental sensations?"

Correct. I use the term "mental sensations" just to imply that they're not emotions, like anger or frustration; they are sensations that feel like cognitive processes.

For example, if you go to bed at night and can't remember the name of a friend, and you say, *I'm going to try to remember his name.* Then, you wake up in the morning, and it feels like the name occurred to you because you have no sensation of actually thinking. You have a sensation only of the idea or the answer coming to your mind in the morning. Then you give that the label of, *Aha, it occurred to me*, as opposed to, *I thought it*.

On the other hand, if you are thinking about something and you get an idea, you say, *I thought that idea consciously*. But what you really sense is a sense of effort and a sense of thinking. The key idea is that there is no reason to believe that the calculation or the computation of the unconscious cognitive process that causes it is any different-whether or not you've had the feeling of doing it.

So, the difference between conscious and unconscious thought, in that sense, would be the difference between how you feel about it. These collectively will be the mental sensations which drive our understanding of the mind.

 My impression was that you really had two themes in the book: that's kind of the underlying science part, and the other one is you spend some time critically examining some of the recent trends in neuroscience.

Correct. Two parts of the book are welded together: the first part really is to complete the idea of mental sensations and the second part is to show how these actually inform the way that scientists think about the mind.

If you have a very strong sense of causation, then you might have a much stronger idea about how *A* causes *B* than someone who doesn't have that same feeling. Therefore, you might come up with an entirely different view of whether or not, for example, a gene causes specific behavior. In other words, what I try to point out is the limitations of neuroscience based on the philosophy of science and also how the brain creates thoughts in the first place.

So, even from the very beginning of the book, you start by pointing out the inherent limitation of looking to neuroscience as our main tool for understanding what it means to be human. I mean isn't that something you're trying to criticize—or at least tell us to stop and think about?

There's no question that neuroscience is really good at furthering our understanding of how the brain works. But the brain is different than the mind, and it's certainly different from being human.

What I really want to do is show that there are certain limits in a mind that is generated by a brain—with all its involuntary mechanisms— trying to study itself. Particularly, there should be a certain degree of humility in how a neuroscientist will interpret the data in ways that are beyond the scope of what is legitimate. (When I use the word "neuroscientist," I mean in the broad sense—cognitive psychology to basic bench science.)

There is a legitimate understanding of how a neuron fires, but nobody has the slightest idea of what a mind is. You can argue endlessly about how neurons fire, and you can do beautiful studies, but you can't extrapolate that into a scientific explanation of the mind. However, you can use it to give you some working idea of how a mind *might* work. Having bits of information is actually different from drawing conclusions.

That is why I argue in the book that neuroscientists should be aware of their limitations: how they came up with their ideas, how they proved them, and what their conclusions will mean as we go forward.

 I guess anybody who claims that they are studying from a totally objective point of view is ignoring the evidence.

That's right. I guess part of the reason I've written this is that there are zillions of books out now about how we are basically filled with biases and all the rest of it; even though we have all that understanding, we still have a suspicion that there is some residual element of pure reason that can overcome these biases. But that's because our brain tells us that, rather than it actually being true. This is just one more nail in the coffin of rationality, so to speak.

But you also emphasize that science is still the best tool that we have for trying to understand how the brain makes the mind; but it has severe limitations.

Right. I want to make it clear that science is the only way for understanding the facts of the world. I'm not a believer in some supernatural things, or the rest of it; I'm a super skeptic.

I mean science is the only way—hard empirical evidence and supporting theories—the only way that you can understand the brain. But there are limitations in making a leap of a purely physical study of the brain to describing something which occurs at another level of existence—namely, the mind.

The mind isn't a physical object that can be studied scientifically. One can infer what it might be by looking at basic biology, but that inference is really just a concept that people have. And there are as many different ideas about the mind as there are people that are alive, probably.

From talking to so many guests I have learned that everyone seems to have their own definitions of the mind and consciousness. I guess this just shows how abstract these concepts are. But one of the implications of recognizing the limitations of science here has got to be that this is not a problem we're going to solve by having better technology.

That's right. What I've tried to present in the book are a number of thought experiments that show the limits of what you might know. For example, if you knew in every instant of a person's life what every cell, synapse, and neurotransmitter was doing and if you believed that neuroscience could unravel the mind, you would have to believe that if we knew all of this we would have a one-to-one correspondence with the mind.

Technically that would be extraordinarily difficult, but it would be impossible to prove or disprove. For example, if someone has a memory and that memory is lost from recollection, we have no way of ascertaining whether that's part of his brain status going forward. Because if he can't access that memory to tell us (and we know how faulty memory is in general), then we can't really have a complete picture.

For example, they've got this Human Connectome Project now. Sebastian Seung, who has written rather eloquently about this, says that if we knew every synapse, we would know the pattern of our thoughts. But the problem is who would tell you the thought?

 Especially if they get that wiring diagram from a dead brain.

Yes. And he actually goes on to say that if you didn't have a post mortem artifact, theoretically in the future you could freeze a brain, and if you had the perfect wiring, you could re-create the thought. Keep in mind, the only way we know what a thought is, is by what people tell us, and we have very little access to how our thoughts originate.

It is fascinating that people talk about rationality and irrationality as though they're whatever we experience. But the truth of the matter is, nobody knows what a thought is. We don't even know if a thought occurs sequentially in a line of reasoning or occurs all at once as in parallel processing and you get a conscious gestalt. Nobody knows what a thought is. So, I don't know how you'd know when you had a thought. I don't know how you'd know how it occurred. There's no way to measure it-except by talking to a person, and there's no way to do that.

There's a fundamental limitation also about the nature of intention. If you try to remember Joe Blow's name and you can't remember it, then you remember it two weeks later, you still intended to do that for two weeks, but you didn't intend to do it consciously. It wasn't part of your awareness, because the intention has been transmitted into your unconscious calculations, where it chugs along until it gets you an answer.

I don't how you'd possibly know how to do a study that would determine unconscious intention. It just makes no sense. So, there are some fundamental philosophic limitations to studying the brain and trying to extrapolate to the next level up, which is a mind.

We'll come back to intention again in a little while, but I would like to explore this is the idea of the mental sensation. Would you define what you mean when you say "mental sensation?"

It's sort of a term I've coined that I'm not entirely happy with, but it's the best I can come up with. If you imagine emotions: everybody knows what anger feels like and most people know what feeling states are, like embarrassment, pride, humiliation, or whatever. But there are some sensations which feel like thoughts. For example—going back to the other book—if you conclude *I'm right*, that's a feeling. But it's a feeling as though you made a cognitive choice. It's feeling that feels like cognition, even though it's an involuntary sensation.

In contrast, for example, if you stub your toe, you don't go through some complicated thought process, you merely say *that stubbed toe causes pain*. You have a sense of causation, which is based upon the proximity in terms of space (in other words, you hit the toe and the pain is in the toe), and in time (it occurs right afterwards), and experience tells us that the stubbed toe causes pain.

But the sensation of pain is a physical sensation. The sense that the stubbed toe caused it is not a conclusion you draw, because you know that even before you have a chance to think about it—instantaneously, *Oh, that's why it hurts*. So, the sense of causation that you get with things like that will then extend to thoughts.

First you have thought *A*, then you have thought *B*. You believe that thoughts can have physical effects, or at least can change things—if you believe in any concept of volition and agency, you believe that thought *A* can cause thought *B*. Well, if thought *B* follows thought *A* in close proximity—just the same way as pain follows a stubbed toe—you eventually get a sense that thought *A* caused thought *B*.

In effect, you're having a conclusion that your brain is deriving—this is not a new idea. This has come from David Hume—you have a conclusion that thought *A* caused thought *B* (because of the temporal relationship and the sense of agency), and you therefore believe that you have a line of reasoning. But, in fact, the feeling of causation that you had was simply just that—a feeling.[29]

So, the main mental sensations I discuss in the book are agency and causation, and of course certainty, which we discussed last time. There are even things like a sense of choice—a sense of whether or not a decision you made was a choice that you made.

There's a sense of effort. For example, when you try really hard to do an algebra problem or something, you have a feeling you're trying really hard. Well, your brain is just doing whatever it does. It might interpret that it's hard; therefore, it gives you a sense that greater effort is required. So, the sense of effort, in terms of even doing something cognitive, would be a mental sensation. So, those are some of the main ones.

I want to take a moment to emphasize one of your key ideas, which is that the origin of these mental sensation is both unconscious and involuntary. It is not anything we have any control over. The same thing is true for feeling of knowing that we talked about earlier.

You also made the point that, in a sense, a thought has two parts: it has the computation part going on in the brain, and then the second part, which is our mental sensation that goes with it. But, if that mental sensation is missing, we think the thought popped out of nowhere.

---

[29] David Hume, *An Enquiry Concerning Human Understanding* (London: 1748).

CHAPTER 15

MENTAL SENSATIONS AND THE SENSE OF SELF

 So, I was just wondering, is our sense of self a mental sensation, or is it made up of these others?

In the first chapter of my book, I describe aspects of a sense of self. For example, a typical out-of-body experience, which you can actually reproduce by stimulating the right temporoparietal lobe in a certain region, is possible, and people can have a sensation of floating above their own bodies.[30]

Now, in that, they have a physical sense of the dimension of self; namely, they are not where their body is located, but their "self" is hovering above the physical body. You can actually do experiments where you can project your sense of self out into space, onto a manikin, or something, through some visual tricks.

So, in that sense, you have a physical sense of a presence, which is sort of the most rudimentary aspect of a self. Then, you have a sense of a self

---

[30] The generation of out-of-body experiences was discussed in great detail in our interview with Thomas Metzinger, author of *The Ego Tunnel* (New York: Basic Books, 2009); see episode 67 of the *Brain Science* podcast at https://brainsciencepodcast.com. Outside the recorded interviewed, Dr. Burton expressed great respect for Metzinger's work.

that is created by agency, which would be an involuntary mental sensation. All of these collectively will give you a sense of a minimal self, by that I mean the ingredients of a self, onto which you then paper the walls with all your memories and stories about yourself and everything else.

So, the complex sense of self that you have is probably a combination of the involuntary sensations that give you a rudimentary sense of self, plus all of your memories and stories you tell yourself and others about yourself. A fully developed sense of self is a combination.

SENSE OF AGENCY

One of the mental sensations that you spend quite a bit of time with in the book is this sense of agency. What do we know about how the brain creates this sense of agency that we expect in our normal day-to-day life?

I think the easiest way for me to understand it is if you think about how you learned, let's say to play the piano. On day one, you can't do anything, and you gradually learn to lay down "Chopsticks." After a period of time, it becomes second nature.

What has happened is you lay down what they call a "predictive map," in other words a template of what "Chopsticks" sounds like, how it feels to play it, in your brain; then, you automatically fire it up whenever someone says, "Play 'Chopsticks.'" In that sense, you have a predictive map in your brain for behaviors that are going to take place. If it happens as expected, you have a sense that you did it.

So, if you were to think about recognition of a face, which, in fact, is the firing up of the predictions that it is a particular face, and it, in fact, turns out to be that face, you also get a sense of recognition. If you realize that you have a pattern of prediction for a thought or for an action—such as playing "Chopsticks"—you get a sense also that you're doing it, and that is the sense of agency.

So, the sense of agency is actually one of those mental sensations that comes into consciousness along with the prediction that created the action. It is yet another way of saying, "Yes, I did that" because it would make no sense, from an evolutionary standpoint, to say, "I didn't do that." You'd never learn anything, and you would be completely paralyzed.

So, you need, from an evolutionary standpoint, a sense, "Yes, that's me doing it." And that "Yes, that's me doing it" is to an action as "Yes, that's Joe Blow" is to the concept of recognition. So, to me, they're all the same thing; they're mental sensations that come along with a preexisting map that tells you that things are going as expected.

 So, timing is critical.

 Yes.

 And people like Daniel Wegner, who manipulate people's sense of agency in the lab, do that by screwing up timing?

 Exactly.

 So, what is the relationship between agency and intention? You told me before that we could actually have intention and have it be unconscious.

 So, you have to ask yourself what you mean by "intention." Let's go back to remembering the name. Let's say I run into somebody, but I can't think of their name, and I say to myself, "I wish to remember their name." Well, that's a conscious intention.

But I can't remember it, and I can't remember it, and I know from past experience, if I think about something else, maybe it will—pardon the expression—come to me. Well, that "it will come to me" is me believing my brain will sort it out if I leave it alone. If I don't prod it and poke it and let it do its work by itself, it will come up with the answer.

Well, think about the brain as being somewhat of a computational device. I hate to say it's like a computer because that makes it like a machine, but it is a computational organ of some sort. So, imagine that you had a blank Google search box, and your computer is sitting there quietly. It's not going to do anything until you enter something in the Google search box, say "Joe Blow's name." Once you enter that in there, you no longer have to sit there; it will come up with the answer eventually.

Well, if you think of the brain that way, you are in effect transferring intention from being a conscious effort to an unconscious command to the brain. In fact, there probably isn't any difference. Once you come up with the idea and then try and think of the name, the brain is going to do whatever it's going to do. The only difference is you put it out of your mind because it takes a while for it to sort through.

Philosophers seem to have a different concept of intention, but frankly, I don't understand intentionality in philosophical terminology. For me, and in common sense use, intention is the desire and the intention of doing something. In this particular case, it is embedded unconsciously.

The idea of an unconscious intention doesn't seem to make sense, but that's the way the vast majority of what we do actually takes place. It's beyond science, in the sense that there's no way of knowing about it, but it's what facilitates anything other than very, very short-term actions and thoughts.

You made a really good point in the book about the fact that since our working memory is actually so limited, only the very simplest thoughts could even theoretically be totally conscious; that we really have to have a lot of stuff going on in the unconscious[31] or outside of our conscious awareness, or we really wouldn't be able to get much accomplished.

That's why I mentioned the Google search engine, even though it may not be a completely apt metaphor. You don't have to put much into the Google search box for all the algorithms to take place, and despite

---

[31] I hesitated to use the word *unconscious* here, but it hard to find a better term.

little input, you get the most complex answers; you don't need to calculate pi in your head; you just say "calculate pi," and the next thing you know, it goes to as many digits as you want.

So, the unconscious computational ability is staggering for the brain, but we can only remember at most maybe four chunks of information. If you take numbers, the average person can remember seven numbers; if you do a phone number with an area code, you chunk the area code into one, the prefix into two, and then the four digits into three. That way you can remember ten numbers. Maybe some people can remember nine chunks of information, but beyond that, unless you take savants, we can't do much more.

We can hold a few facts in our brain at once, and if you want to do a thought that involves, let's say ten variables, after you've done four, you have to start putting them into long-term memory in the unconscious, in order to get the others onto your so-called "minimal RAM [random-access memory]" of conscious memory—or working memory—which means that almost all of complex thought, by the very definition, has to occur in the unconscious.

And we don't have any way of telling what parts of what we are thinking came from where, that is, which part is coming from the unconscious and which part is conscious. So, there is really no way we can have anything that we can say, "I know I thought that totally consciously."

That's right. And I actually didn't mention, when you asked me about mental sensations, one of the mental sensations we have is ownership. For example, we have a sense of ownership of our body parts—"mineness."

We also have that same sensation, "that's my thought," because it is a possessive sense that this was my thought that occurred and because I have a sense of agency of a mind that created it. Yet, the vast majority of this work is going to occur out of sight, where it can have influences that are far-reaching, beyond just the individual. Group thought can be occurring. That's why there's so much cultural determination of how we think—much of which we're not even aware of.

CHAPTER 16

AGENCY, CHOICE, AND CAUSATION

 What about the relationship between agency and choice?

 It's hard for me to say because this is obviously just my way of thinking about it as sort of speculation, as opposed to something you can prove. But, if you think about it, agency is the sense that *you* are doing something; for example, you meant to kick the ball or whatever activity you performed. Choice means you meant to make that decision.

So, to me, the sense of choice is really at the heart of the free will question. In other words, a lot of people debate as to whether or not we have free will, and John Searle's argument is, "Well, no; if I go into the restaurant, I know that when I choose a hamburger over tofu, I'm making a choice." What he's really saying is he has a very strong sense of agency for his thoughts.

So, for me, agency for thought is the same thing as a choice. In other words, you feel that you're actually exerting influence with your thought on another thought.

 So, the feeling that we are making a choice is another example of a mental sensation?

Right. All the studies by Ben Libet and all the pros and cons of arguing about free will stem from this very phenomenon that you have a sense of choice. You think that that's when you chose to move your hand or whatever. However, the studies that timed the movement of the hand actually show that the feeling of choice might not coincide with when your brain "decided" to move your hand.

You've got a whole field devoted to those studies.[32]

Yes, I'm not going to try to get into that today, but you did devote a whole chapter to causation. I assume this is a reflection of your current philosophical bent since you said you were trying to look at Hume from the standpoint of what we now know about the neurophysiology. Would you like to talk a little bit about what we know about how the sense of causation arises?

Okay, we'll go back to the example I mentioned, you stub your toe. Let's say you stub your toe, and one instant later you have pain that emanates from your toe. You're pretty sure that *A* caused *B*—that *stubbed toe* caused *pain*—because your brain has learned that *trauma causes pain*.

The first time you touch a hot stove as a little kid, you don't have any idea that it's going to cause pain, but after you've burned your hand once, you know that there's a strong relationship. You know that only experientially; you can't know that from any other thing except through experience. There's no way to prove it without experience. This is where Hume's argument on induction comes from.

So, the first time you do it, you're unaware of it. Then, the second time you say, "Oh, I'm not going to do that; it causes pain." The close proximity in the timing of these events is an important clue. If you stub your toe and the pain starts a month later, it's hard to know whether it caused it, maybe it's something else. Maybe, you've got gout or something.

---

[32] Libet's recorded signals in the brain that occurred before the person decided to move their hand. See the previous discussion in Chapter 10.

So, the closer in proximity temporally, the more likely it is, and the closer it is spatially, the more likely it is. It's really based upon the fact that you believe that *A can* cause *B*. In the David Hume example, he uses billiard balls, and he says, that if you believe that billiard ball *A* can cause billiard ball *B* to move, and you shoot billiard ball *A* at billiard ball *B* and it moves—because of the temporal and spatial relationship plus the belief that the ball can cause another ball to move—you've got all the requirements for causation.

Well, in my book, what I really wanted to point out is, this extends even to things like a thought. If one thought immediately follows another thought and I have a strong feeling that one thought *causes* another thought, I will believe that I am actually exhibiting a line of reasoning that *A* caused *B* caused *C* caused *D* in a series of thoughts that leads to a final conclusion. But, as I mentioned a few minutes ago, we know that you can only have four chunks of memory at any one time, so you couldn't have a ten-idea line of reasoning that occurred *exclusively* consciously, even though it would feel that way.

That sense of causation for thoughts is terribly important to our understanding of what it means when people say, "Oh yes, this is how the brain causes that." Science is both a matter of getting evidence and trying to infer causation, but when causation is inferred in these complex things, such as from brain to mind, it is subject to the same involuntary mechanisms as all mistakes of causation are.

I see causation as the one supervening variable of involuntary thought that plays most havoc with us understanding disease and understanding what causes what and, particularly, understanding how the mind causes behavior and the brain causes the mind. The sense of causality is really an involuntary sensation, rather than a proven fact.

 And, as Hume said, it is based on our experience. So, would you consider causation, in a way, to be a learned mental sensation?

 Yes, I think it is learned from experience.

I remember when we were talking about *On Being Certain*, we talked about the fact that people have innate differences between how prone they are to certainty. I'm wondering if then we might see a similar variability about people's tendency to attribute causation, especially since it's unconscious.

I think of these mental sensations as a particular, or peculiar, depending on how you look at it, kind of traits that everybody will be afflicted with to a greater or lesser degree. For example, some people will feel a great sense of certainty, the so-called know-it-alls, and other people will be sort of doubting Thomases, true skeptics.

When it comes to causation, I think the same must be true. For example, when I see things, I rarely conclude causation; I conclude association or correlation, but I rarely have a feeling of causation. I'm always surprised when people say that "this causes that." On the other hand, other people are absolutely convinced that *A* causes *B*, and I think in part it may be an inherent difference in the degree to which it is expressed, just like you said.

I suspect that those people that go out on a limb stating causal mechanisms, when you're bridging the brain/mind gap, are particularly striking and variable. When people think we can explain everything about the brain, and therefore, understand the mind, I suspect they have a much stronger sense of causation and agency than those that think we'll never get there. So, it's part of the difference between doubt and belief and alternate understanding; it may, in fact, be part of these mental sensations.

I'm going to quote from your book.[33] I liked it when you said, "*One investigator's possible correlation is another's absolute causation.*" We see that in action all the time, and it's actually also one of the problems when

---

[33] Unless otherwise indicated, all quotes in Part 3 are from *A Skeptic's Guide to the Mind* (New York: St. Martin's Press, 2013). Page numbers were not included because I was working from an exam copy, and I remain unsure about the page numbers of the final version.

scientific results get out into the press; correlations are constantly being misquoted as evidence of causation.

 Right. And, to me, the only value of my ideas is if you recognize that causation itself is a sensation.

Now, let's take something that most people would agree on: that cigarette smoking causes lung cancer. Well, you can get an enormous amount of evidence to support that position; both epidemiologically and with lab tests. You can get a pretty high degree of likelihood that cigarettes can cause lung cancer. It's really when you get into these more complex issues about whether, let's say, a disorder in the brain is responsible for loss of libido, or an fMRI scan shows x or y, and you draw a conclusion.

When you get into these complex issues, the sense of causation is almost all that people have to go on in determining, "Oh yes, this data is statistically significant, but it doesn't indicate definite causality" or "This is statistically significant, it must be that A caused B." The conclusion that arises out of the statistics is purely that involuntary sense; the statistics stand on their own. That's really hard to explain to people.

The irony is that the stronger a person's individual involuntary sense of things like self, causation, and certainty, the greater their belief is that the mind can explain itself.

 That's right.

It made me think about the contrasting personalities of Descartes and Hume. I mean those are the two ends of the spectrum right there.[34]

 Right. That is exactly correct.

---

[34] Hume believed our knowledge is based on experience whereas Descartes believed that all problems can be solved by mental reflection and logical thinking.

And the thing is, if I have that feeling—if I know that I didn't spend any time thinking of a guy's name and it suddenly occurs to me—and I say, "I wasn't thinking about it," there's no way someone is going to convince me that I was thinking about it because I have no experience of that. So, if you have an experience of causation, which is strong, nobody can really argue with you.

This is one reason why you see so much alternative medicine. When someone tells you *x*, *y*, *z* works, but you say, "No, there's no empirical evidence." They respond, "Well, I know, because I did it and I felt better." Basically, what they're talking about is the sense of causation.

This reminds me of another one of my favorite quotes from your book, "Hiring the Mind as a consultant for understanding the Mind is the metaphoric equivalent of asking a known con man for his self-appraisal and letter of reference." That's kind of where we are.

Yes. You know, I've spent ten years on these two books. It is humbling to realize how little we are in control of our thoughts and how much of these sensations of a mind are, in fact, the vagaries of an unconscious brain telling us, according to its own agenda. It is very humbling.

We don't have any way out of this paradox, and I'm going to quote you again, because you said, "The less-than-perfectly-reliable mind will always be both the mind's principal investigator and tool for investigation." It's as if we were trying to use a microscope to study a microscope; we're kind of stuck!

Yes, we are stuck. It's fascinating because I know you've done all these podcasts with people, many of whom would agree, in general, with this principle, but at the same time, there's a spectrum of how much people think they can overcome this basic paradox through more brute work. I know Daniel Dennett, when he was asked about this, he said,

"We'll just have to try harder." Yes, I understand that, but that won't get you there.

It's going to be interesting to see what Sebastian Seung's position on this is fifty years from now—whether or not we have his wiring diagram.[35]

Well, you know what's fascinating about him is, I did a little background check, his dad is a professional philosopher, and he's very well-educated. I actually think he studied philosophy in college.

It's clear to me that this is a strong sense that you can uncover causation, at this level, and the causal mechanisms. I wouldn't be at all surprised if people, like Seung, down the road, will be discovered to have a very strong sense of causation that drives these statements about causation. There's a sort of circularity to the whole thing.

It would be kind of amusing if fifty years from now you went to the Human Genome lab and they said, "You really have a strong sense of causation," and then the guy says, "What causes that?" And so on. I'm just making a joke, but you know what I mean. I think a lot of the issues in neuroscience actually emanate from these different involuntary positions generated by these sensations. That's one reason that I wrote this book.

So, I think we should want to understand these limitations and realize that, since the mind really is the ultimate con man, we have to be on constant guard against our built-in tendency to overestimate what we can do.

Well, that's the goal. But, even as I'm talking to you, I have some idea that my ideas seem like they might be correct, or I wouldn't be

---

[35] Sebastian Seung, author of *Connectome: How the Brain's Wiring Makes Us Who We Are* (Boston: Houghton Mifflin, 2012), was interviewed in episode 75 of the *Brain Science* podcast, available at https://brainsciencepodcast.com.

telling them to you. It's hard to shake your own beliefs. I will say, having worked on this for ten years—the two books—it does make you always say, "Yeah, but how do I know that?" You have to work at it full-time. It's tough not to just become prey to your own thoughts.

 As you said, the real take-home message is humility. And that brings me to an example that you brought up in the book that I think is worth visiting, which is this Dunning-Kruger effect. Would you talk about that for a minute?

Oh, yes. Dunning and Kruger did this study[36], which is fairly well-known by now. Basically, they took a group of Cornell undergraduates and looked at self-appraisal in terms of various intellectual tests.

I was most interested in the one about making logical inferences, because it turned out that the people who did the worst estimated themselves as being in like the 68th percentile, in terms of overall how they thought they did compared to the group—even though they were in the bottom 12 percent.

They concluded that those people with this inability to assess their own cognitive abilities don't know it. They uniformly tend to overestimate their own cognitive abilities. So, there's a double impairment: the inability to do so results in a faulty assessment of your ability to do it. We see this everywhere.

Now, did they have any theory of why this was?

Not really. I actually wrote to them. I think what it amounts to is that, in order to know that you may be wrong, you have to not have a very strong sense that you're right.

---

[36] J. Kruger and D. Dunning, "Unskilled and Unaware of It: How Difficulties in Recognizing One's Own Incompetence Lead to Inflated Self-Assessments," *Journal of Personality and Social Psychology* 77, no. 6 (1999): 1121–34.

It seems to me that, if you don't know you're wrong, you're not going to get that sense of unfamiliarity or uncertainty that other people do when they are asked a question when they're not sure of the answer. Perhaps, when people don't know that they're wrong, they lack the facility to recognize that disquiet or discomfort that comes from a wrong answer.

So, when you see people arguing about stigmata, or the 6,000-year-old universe, or something like that, you go, "You've got to know that that's wrong." But no, they don't have to know that that it is wrong. If they have a strong enough feeling that it's right, you can't convince them otherwise.

A large portion of what we consider irrational thought actually is predicated upon an inappropriately hypertrophied sense of rightness. Now, how that comes about, whether that's a biological trait people are born with or whether it's acquired as a defense mechanism or whether it's a combination, I don't know. Maybe that's something for future research, but I think that, to me, explains it to some degree.

CHAPTER 17

THE FEELING OF FAMILIARITY

Now let's explore a related phenomenon. It has been consistently shown that when people are given tests, they are more likely to pick the thing that's familiar as the answer. Will you talk about that?

If you think about familiarity, it's a way of your brain telling you that you're on the right track. So, if you were to take familiarity as a cognitive sensation, or a mental sensation, it's just like when you take a multiple-choice exam; you may not know the right answer, but one feels more right than another. That one tends to be—although it's not necessarily so—right, more than the others. So, familiarity is a way of us sensing that our thoughts might be correct, even if we don't know that for sure.

Now, the speed with which you process information will give you a greater or lesser sense of familiarity. In other words, the faster you can process something, the more normal it feels—therefore, familiar. This goes back to the prediction model I mentioned earlier. If you have in your mental map something, an idea, and this closely approximates it at first glance, you will have a sense of familiarity. If it doesn't approximate it, you won't have a sense of familiarity; therefore, it will take longer to process. You'll have to work it through with more effort, and it will seem less likely to be correct.

So, those people who are really smart, who process really quickly because they're a so-called expert in an area, they're likely to make a snap judgment because it feels right, right away. They're also likely to say, "Ah, that's right." The faster they do it, the more likely they are to feel that they're correct, but they may also immediately conclude "That's wrong; I already know that." It's another way in which the real smart people can be lulled into this sense of being right, simply because of the speed at which they process information.

 I was wondering if this relates to the popular idea that the first thought that comes to your mind is a good one and that you should trust it as somehow special, just as intuition is often regarded as unusually reliable. But that isn't necessarily so. However, it's easy to understand evolutionarily why we would be this way because snap judgments would have been really good for avoiding being eaten.

 Right.

 But maybe not so good for the complicated world we live in now.

 Well, you know I think that's probably true, and I think that if you think of the brain evolving piecemeal and for different evolutionary reasons over the millennia—or longer—you could see where a function that would work well in a simple society might not work well in a complex society. In the book, I use the idea of global warming. It is impossible to take in all the variables. Snap judgments cannot work on something like that.

Snap judgments work best when life is being threatened—immediately being threatened—but they don't work best when life is being threatened at another level, like with climate change or with the Iraq war or whatever. We've moved to a new era, where these primitive sensations need to be recognized for what they are.

And, even if we think we're smart, that must mean we need to be even more careful.

 That's right, and actually we're not very smart. If you think about how far we've gone, we're still debating 2,500-year-old ideas in philosophy. Though neuroscience has been fantastic, nobody knows what a mind is. Nobody knows what a thought is. We have no idea.

So, we have really smart people working on it—I'm not one of them—but you have some really smart people working it. However, it's sort of like standing and looking at the stars and thinking about your own life. We have wonderful lives, filled with meaning, and we look out at the stars and go, "Oh, my God!" Right?

There's a long distance between us and anything, and I think the same goes for science. So, smartness, to me, is sort of a short-sighted, egocentric view of the world. Excuse me for saying that.

 I understand what you're trying to say.

SUMMING UP DR. BURTON'S SECOND INTERVIEW

As usual, we've left out a lot of information, but I would like to take a few minutes to go through the principles that you outlined at the end of your book. I agree that these are good principles to remember when we read neuroscience claims about the mind and maybe even science claims in general.

Right. If there were only one message that I would want people to go away with, it is that the Mind isn't a specific organ—it's not a liver, it's not a spleen, it's not an eyeball—it really exists in two dimensions. One where we experience mental sensations and one that we experience as a mind.

The first part of the book really is designed to show how we experience these mental sensations, which collectively make up most, if not all, of the mind. I don't really know; I can't wrap my mind around my mind, so, it exists primarily in how we feel about things: we feel we're thinking, we feel we're deciding, we feel we're choosing, we feel we're acting in a causal manner, et cetera. That's how we experience it.

However, we also talk about it in the abstract. For some neuroscientists the mind would be any cognitive process, even if it's unconscious, and for other people, it would extend to include group minds or group behavior or even the swarming behavior that we see in animals.

But the first-person experience of a mind is not something that you can study with a scientific inquiry; it's really based upon what people describe, and given the fact of memories and exact descriptions and given the problems of language and attention, we will never get a complete picture of a mind from a person.

If you think of James Joyce and *Ulysses* or *Finnegan's Wake* or Virginia Woolf's *A Room of One's Own*, those are examples of a stream of consciousness, which is really smart people trying to give you a picture of a mind in action. It gives you a sense of it, but it doesn't give you the completeness of it, nor do you get any idea of how those thoughts came about.

This second abstract concept of the mind drives most of modern neuroscience. Many of these differences boil down to how you think about what a mind is, but since a mind is a concept, then all concepts are equally valid to the extent that they can be supported by reasonable scientific evidence. But you could make arguments for an extended mind, a group mind, an individual mind, a mind for organisms that are exclusively unconscious, and so on. Literally, the sky is the limit.

Thus, when you read about the neuroscience of the mind, you have to understand you're reading about a concept. It's not like cancer that you can look at under a microscope; it is a concept. It exists as experiencing a concept. In fact, if I could get that one point across, the book would be worth it.

The other point I want to make is that all thoughts about and studies of the mind are guided by involuntary brain mechanisms that collectively generate an illusory sense of a personal, unique self, capable of willful, unbiased exploration of how a brain creates a mind. I realize that I have a very strong sense of self, and to some extent, I'm proud of it. On the other hand, I recognize a lot of it is just biologically determined; good luck in whatever genes I have.

The sense of self is not what we think it is; most of it is just driven by our biology. The narrative that we have about it comes out of this biology too—in the sense that you do well in school, and someone says, "He's a smart kid"; then, you think you're smart. Then you tell other people you're smart, and so it goes. A lot of this may be beyond your control. I would like people to understand that scientists looking at the mind are operating primarily through the mixing and matching of these very sensations that drive their own investigations.

The third key idea is that taking into consideration how these mental states create a sense of mind is a necessary first step to a real, albeit partial, understanding of what a mind *might* be. Failure to acknowledge these biologically imposed limits on a mind examining itself will only result in further neuroscientific excesses.

I think those are my main take away ideas.

 That's a good summing up. And just remembering that this paradox is not one that we're going to escape by having more technology or better technology; it's going to be with us as long as we're trying to study this.

That's right. You've summarized it better than I could.

 As I said, I know we've only scratched the surface of this, and it's really an important topic. Is there anything else you want to share before we close?

Well, I do want people to understand that my book is not a criticism of neuroscience or the scientific method because, obviously, science is the only way of understanding the real world.

I just want to emphasize that *the tool for studying the mind ultimately is still going to be the mind, and the mind is not totally accessible to critical analysis. So, the main tool for studying the mind is essentially always going to be incomplete and subject to the various errors that are built into the biologic machinery of our brain.*

So, if we were going to be good scientists, good scientists always recognize the limitations of their techniques and methods, so recognizing this problem would be an example of doing what we should be doing anyway.

Right. I feel really strongly about this. Data is data, but the interpretation of data is story. If I find an fMRI scan shows *A*, *B*, or *C*, and I say, "This is an example of how this patient was or wasn't conscious," or whatever, what I'm doing is I'm extrapolating the data into a conclusion.

Conclusions are stories. Now, if the data is perfectly accurate and the scientific method is great, then there's a high likelihood that that story is accurate. But the scientist needs to understand that he's telling stories.

Ideally, if I had the political energy, I would want science of the mind journals to require that everybody who writes a paper include a short biography and reason why they wrote their paper. Now, maybe they'll have no self-awareness whatsoever, but you at least will get a clue as to what they were thinking.

It's folly to think that science of the mind is exclusively pure science. Neuroscience of the brain is science, to the extent that you can eliminate the scientific error, but the mind is not scientific inquiry of the same nature, and people should be forced to say why they think the way they think.

I think if science would recognize that it's not the last word—it is another word when it's applied to the mind—it may be the best data, but it still is filtered through all these biases. And I know that sounds like a sermon, but …

 It's a sermon that we need.
Do you have any advice for students?

If I were a student today, I would think that we're on the cusp of a new era. If you think about it, I'm at the age now, when I was growing up people studied Freud and James Joyce. The unconscious was a boiling pot of repressed desires, et cetera. It was primarily folklore psychology.

In the last fifty years, we've moved from folklore psychology to an increasing reliance on scientific evidence.

But if you think about it, it doesn't make any difference what science you do on behavior. You have to describe it in folklore terms. In other words, if you say, "I believe this person is a psychopath because his fMRI shows this," or whatever, the word "psychopath" is a psychological term, it is not a physical term, as in a "neuron."

I think that if people recognize that most theories about the mind are not driven by underlying scientific method, as it applies to the biology, but that they are driven by stories that drive the scientist. Then we would understand the concept that we can't truly study the mind. If I were going to go into neuroscience or one of those areas, I would spend a lot of time in the humanities and get the broadest possible view of other ways of thinking about these problems.

You can look at a Chuck Close painting, and you can understand the difference between understanding individual neurons and understanding the mind—by just stepping close to the picture, where you see the pixels, or stepping back, where you see those individual pixels morph into a portrait. You can see that you're working at different levels. But if you don't know who Chuck Close is, you're less likely to have that experience.

So, I would go back to the well-rounded Renaissance person; if you're going to go into neuroscience or science of the mind or philosophy, then take as much in the humanities as is possible.

 Well, I think that this book is a valuable contribution to the conversation. So, I appreciate that you put the years into it.

Oh yes, it's a lot of fun, and it's helped me on a personal level because I think the thing that I've learned after ten years of doing these two books is when you can and when you cannot make a difference interfering in either public life, as in politics, or private life, as in interpersonal relationships. I used to think that there was a single line of reasoning; that if you accepted that, that everybody else should accept it too because it seemed like it was inescapable. Now I recognize that's simply not true, even though it might feel that way to me.

I've got some very strong opinions on things like global warming, but I also recognize that if you stand back from your feelings, there are alternative ways of thinking about it. I'm not saying they're right, but I can see where alternative ways would be appealing. It's stopped me from being quite so anxious to give advice, be dogmatic, or interfere in other people's lives. Doing these books has been like doing a peculiar form of psychoanalysis.

 The years I have spent doing my podcasts has also changed how I see the world around me.

For example, when I'm dealing with my patients, I'm very aware of the nonverbal communication that's going on between, for example, parents and their children. And sometimes, I'm painfully aware of the fact that the patient doesn't seem to live on the same planet as I do, but I don't always know how to overcome it—in fact, I usually don't know how to overcome it, but I'm more aware of it than I think I would be. Plus, I feel it's much better than psychotherapy.

I think in my last book I mentioned how the brain creates a sense of purpose and meaning. Recently I was listening to a series of tapes on existentialism, when something stopped me cold. It was a guy saying, *Well, you know, everybody has a different sense of purpose, and for some people it's happiness or virtue or whatever.* He said, *With some people it's retribution.*

I've thought about modern American politics recently, and I thought, "Why can't these guys come to a reasonable compromise on issues that affect everybody?" And everybody is arguing as though if we could compromise, we'd solve the problem.

Then I realized, they don't want to compromise, but not because they may not want to compromise, it's because their primary pleasure in doing this is to generate a sense of purpose in their lives that might arise out of some negative emotional attribute, such as vengeance, retribution, or just ventilation of anger.

And I thought, "Gee, what if that really is their purpose to life?" Well, that's quite different than thinking they're trying to be reasonable and rational. Rationality takes second place to the feeling of whatever gives them real meaning. I mean this is true on all sides; I don't mean to single out one group of people.

It's allowed me to understand that these mechanisms—these mental sensations—probably drive the majority of modern discourse; it's not reason. That's really an alarming and discouraging thought, but I think it's true. How about that for a piece of pessimism?

 Well, I wish I could think of something more positive to end with. You left me almost—almost—speechless.

But to emphasize this, most of modern discourse is not about arriving at the best answer; it's about arriving at the answer that gives the individual participants the greatest sense of pleasure and purpose. That's quite different, and until you can see that—I don't mean you personally—but until you recognize that, then we're sort of stuck.

So, all we can do is put out our little piece of information that's as accurate as we're able to make it and hope that the ripples eventually go someplace good.

That's the way I look at it. You never know what contribution is going to turn out to be valuable.

I cannot begin to tell you how many emails I've gotten from the last book that have been provocative; some have made me think about things differently. To give you an idea, in the same day I got an invitation to speak at an atheist convention and a Southern Baptist University commencement exercise. And I thought, *"How cool is that!"*

So, people do listen to your podcast. I mean I've gotten letters, actually, from people that have heard you. It serves a purpose. You just never know who's going to carry the ball forward.

 As a physician, don't you find it a little bit surprising that you might have made more impact with your writing than you did when you took care of patients?

 You know, my wife has collected, because I'm an egomaniac but not that bad, the printed-out emails. Actually, they've filled up a box. I can't tell you how many people said that just the idea of knowing how certainty works has changed their lives.

As a neurologist I did not get a large number of letters from thankful patients, because I didn't do much. I did the best I could, but a lot of diseases aren't treatable.

Yes, it's been staggeringly rewarding in a funny sort of way. But at the same time, I recognize that I don't know how much of this is original or, just my way of saying, something that is common. It's hard to know how to put anything you do into a larger perspective.

 Well, I have enjoyed talking to you. We need to make it a point not to make it so far apart in the future.

 Absolutely!

CHAPTER 18

SECOND INTERVIEW SUMMARY

First, I want to thank Robert Burton again for taking the time to talk with me. Even though we have never met in person, I consider him both a friend and a mentor. As I was editing this interview, one thing that struck me was Robert's humility.

I strongly believe that everyone should read *A Skeptic's Guide to the Mind: What Neuroscience Can and Cannot Tell Us About Ourselves.* As Robert mentioned, this book really has two main parts. In the first several chapters, he extends the principles he developed in *On Being Certain* to other mental sensations. We tend to take things, like our feeling of certainty, agency, and causation, for granted, but he points out that these are generated in parts of the brain that we can neither access nor control.

But what makes this book stand out is that he explores the implications of this reality. He argues that while we can become ever more knowledgeable about how our brains work, the mind—which is something that we each experience subjectively—is much more elusive. No wonder everyone seems to have their own definition of *mind* and *consciousness*.

Let's review the principles that Dr. Burton says we should remember when we are reading any neuroscientific claim about the mind. These are from near the end of his book:

- The mind exists in two separate dimensions, as felt experience and as an abstract concept. Neither is fully accessible to traditional scientific inquiry.
- All thoughts about and studies of the mind are guided by involuntary brain mechanisms that collectively generate an illusory sense of a personal, unique self, capable of willful, unbiased exploration of how a brain creates a mind.
- Taking into consideration how these involuntary mental states create our sense of a mind is a necessary first step to any real, albeit partial, understanding of what a mind might be.
- Failure to acknowledge the biologically imposed limits on a mind, examining itself, will only result in further neuroscientific excesses.

The bottom line is the fact that we're trying to study the mind with a mind that has inherent limitations. And I think Robert is right when he says our response should be humility.

FINAL REMARKS

In reviewing the transcript of episode 96 of the *Brain Science* podcast while creating the second edition of *Are You Sure?*, I realized that I do not totally share Dr. Burton's skepticism about our ability to study the mind. While I agree that there are certain built in limitations, neuroscientific attitudes toward subjective experience have been evolving. As Stanislas Dehaene noted in his book *Consciousness and the Brain*, the neuroscientific study of consciousness requires that subjective experience be regarded as a form of raw data. This means that, just as for any other sort of data, we must be aware of potential biases and other sources of error.

At a deeper level, I think Dr. Burton was arguing that science, including neuroscience, is not the only tool for understanding what makes our lives meaningful. There is more to having a mind than having a brain. As other writers have emphasized, having a mind is an embodied experience that also involves our interaction with the world, including other living beings, both human and nonhuman.

An example of this principle is the fact that addiction cannot be fully understood if it is regarded purely as a brain disease, due to the fact that this ignores the psychosocial factors that often underlie self-destructive behavior.

Another goal of Burton's *The Skeptics Guide to the Mind* was to argue for humility, both on the part of science and on the part of individuals. To me, this is a valuable take home message. In *On Being Certain*, we learned that our sense of certainty, which he called the feeling of knowing, is not always reliable. It is generated by unconscious processes in our brain, processes that evolved to help us survive in a world of ambiguity and uncertainty. It was better to flee a non-existent tiger than it was to stand still and get eaten.

Over the years since I talked with Dr. Burton, evidence has mounted showing that much of what our brain does is beyond our conscious access, but the practical significance of this discovery is seldom explored. Dr. Burton's writing begins to fill that void.

In *The Skeptics Guide to the Mind* Burton includes the feeling of certainty under the broader concept of mental sensations. The use of the term *sensations* reminds us that these processes are not under our conscious control any more than the sensations associated with perception. Included under the banner of mental sensations are our sense of agency and the closely related concept of causation. What all these mental sensations have in common is that all we experience is the final result, which is to say the conclusions passed on to our consciousness. Just as we cannot control visual illusions, we cannot control whether theses sensations are accurate.

That's why I agree with Burton that our response should be humility. I also hope that this knowledge will help us become more tolerant toward ourselves and others.

Finally, don't forget the concept that Burton emphasized in *On Being Certain*, there is no such thing as autonomous rationality because even our most sincere attempts to reason rationally are influenced by unconscious processes that we can neither access nor control.

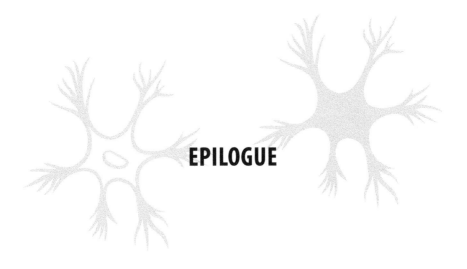

EPILOGUE

These interviews were originally recorded in 2008 and 2013, respectively, but the reason I wanted to republish *Are You Sure? The Unconscious Origins of Certainty* now, in early 2020, is that I feel Dr. Burton's ideas are as relevant as ever. Neuroscience has continued to amass evidence that most of what our brain does is unconscious and inaccessible to reflection. We have also learned much more about how brains generate minds and come to appreciate that consciousness[37] is probably more widespread than we ever imagined.

Unfortunately, public discourse has become more polarized than ever, especially here in the United States. The topics Dr. Burton mentioned (oil drilling and global warming) seem tame in light of recent events. The Enlightenment ideal of rational discussion seems to be disappearing and respect for the value of evidence is being overwhelmed in almost daily displays of the power of cognitive dissonance.

Dr. Burton seems understandably discouraged by these trends, and sometimes I share his despair. But I also believe in the power of knowledge, and my hope is that a better understanding of why human beings behave irrationally is a first step toward a more rational future.

---

[37] In 2019 I did a four-part series about the neuroscience of consciousness. It can be found as episodes 160–164 of *Brain Science*, which is available at https://brainsciencepodcast. com (or in your favorite podcasting app).

ABOUT THE AUTHOR

Virginia "Ginger" Campbell, MD, is best known for her long-running podcast *Brain Science*. "The show for everyone who has a brain" was launched in 2006 with the goal of sharing how recent discoveries in neuroscience are unraveling the mystery of how our brain makes us human. *Brain Science* enjoys a diverse worldwide audience because of Dr. Campbell's unique ability to make complex ideas accessible to listeners of all backgrounds. She spent many years working as an emergency physician and now practices palliative medicine at the Veterans Administration Medical Center in Birmingham, Alabama.

Dr. Campbell is a prolific podcaster. In addition to *Brain Science*, she hosts *Books and Ideas* and *Graying Rainbows: Coming Out LGBT+ Later in Life*. She lives with her golden retriever, Rusty, in Birmingham.

All of her podcasts are available for free in your favorite podcasting app and at http://virginiacampbellmd.com. You can also sign up for the free **Brain Science** Newsletter to receive new episode show notes and other updates.

WHY NEUROSCIENCE MATTERS: COMING IN EARLY 2021

re You Sure? The Unconscious Origins of Certainty was based on three episodes of the *Brain Science* podcast. Dr. Campbell is working on a longer book that explores more about how recent discoveries in neuroscience are unraveling the mystery of how our brains make us human. Look for *Why Neuroscience Matters* at your favorite bookseller in early 2021.

You can also follow Dr Campbell's author page on Amazon and/or Goodreads for the latest updates.

BRAIN SCIENCE: THE PODCAST FOR EVERYONE WITH A BRAIN

D r. Campbell is a prolific podcaster. Her most well-known show is *Brain Science*, which she calls "the show for everyone who has a brain," because her goal is to make recent discoveries in neuroscience accessible to listeners from all backgrounds. She has interviewed more than 100 scientists since *Brain Science* launched in 2006. Listen for free in your favorite podcasting app.

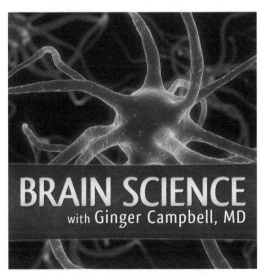

brainsciencepodcast.com

BIBLIOGRAPHY

Bohm, David. *Thought as a System*. Abingdon, UK: Routledge, 1994.

Burton, Robert. *On Being Certain: Believing You Are Right Even When You're Not*. New York: St. Martin's Press, 2008.

Deacon, Terrence W. *Incomplete Nature: How Mind Emerged from Matter*. New York: W. W. Norton & Company, 2012.

Dehaene, Stanislas. *Consciousness and the Brain: Deciphering How the Brain Codes Our Thoughts*, New York: Penguin Books, 2014.

Eagleman, David. *Incognito: The Secret Lives of the Brain*. New York: Vintage Books, 2011.

Festinger, Leon. *A Theory of Cognitive Dissonance*. Stanford: Stanford University, 1957.

Gazzaniga, Michael S. *Who's in Charge?: Free Will and the Science of the Brain*. New York: HarperCollins, 2011.

Gigerenzer, Gerd. *Gut Feelings: The Intelligence of the Unconscious*. New York: Penguin, 2007.

Gladwell, Malcolm. *Blink: The Power of Thinking Without Thinking*. New York: Little, Brown, 2005.

Goleman, Daniel. *Emotional Intelligence: Why It Can Matter More than IQ*. New York: Little, Brown, 1995.

Hood, Bruce. *The Self Illusion: How the Social Brain Creates Identity*. Oxford: Oxford University Press, 2012.

Lakoff, George, and Mark Johnson. *Philosophy in the Flesh: The Embodied Mind and Its Challenge to Western Thought*. New York, Basic Books, 1999.

Libet, Benjamin. *Mind Time*. Cambridge, MA. Harvard University Press, 2004.

Searle, John R. *Mind: A Brief Introduction.* Fundamentals of Philosophy Series. New York: Oxford University Press, 2004.

Schacter, Daniel L. *The Seven Sins of Memory: How the Mind Forgets and Remembers.* New York: Houghton Mifflin, 2001.

Tavris, Carol, and Elliot Aronson. *Mistakes Were Made (But Not by Me): Why We Justify Foolish Beliefs, Bad Decisions, and Hurtful Acts.* Revised and new edition. New York: Houghton Mifflin Harcourt, 2020.

Wilson, Timothy D. *Strangers to Ourselves: Discovering the Adaptive Unconscious.* Cambridge, MA: Belknap Press, 2002.